CAN ANYONE SEE BERMUDA?

MEMORIES OF AN AIRLINE PILOT
(1941–1976)

PUBLISHED BY:
Cirrus Associates (S.W.),
Kington Magna,
Gillingham,
Dorset,
SP8 5EW UK.

ISBN 0 9515598 5 0

PRINTED IN ENGLAND BY:
The Book Factory,
35-37 Queensland Road,
London,
N7 7AH.

PHOTO SCANNING BY:
Castle Graphics Ltd.,
Nunney,
Nr. Frome,
Somerset,
BA11 4LW.

SOLE DISTRIBUTORS:
Cirrus Associates (S.W.),
Kington Magna,
Gillingham,
Dorset,
SP8 5EW.

COVER PHOTOS:
via A.S. Jackson

For Darragh

CONTENTS

CHAPTER 1

TIGER MOTH DAYS

Several thousand feet above the grass airfield of Watchfield I sat strapped into the rear seat of a de Havilland Tiger Moth as the Polish Sergeant Glauber performed a wide and gentle turn, pointing out landmarks clearly visible on that fine early summer day of 1941 – Oxford, Faringdon and Swindon. This was my introduction to flying in the Royal Air Force and Glauber was in no hurry to test my nerve with spins, rolls or other forms of aerobatics. He told me to put my feet on the rudder pedals and to hold the control stick while he demonstrated the effect of the movement of these on the rudder, ailerons and elevators. Then he invited me to try some turns, climbs and descents. I must have performed these reasonably well because I was asked if I had already done any flying.

"No," I replied, "never."

My only previous flight had been some years earlier in a former Handley Page Harrow bomber owned by Sir Alan Cobham, whose flying circus used to tour holiday resorts throughout the British Isles. In addition to exhibitions of stunt flying the public were offered short flights at a cost of five shillings. To the great dismay of my father I had won a free ticket for such a flight when attending a concert on the pier at Littlehampton. His own experience of flying had been gained during the first world war when he had been ordered aloft to observe from a single-engined Royal Flying Corps machine the accuracy of the shooting of his company's guns on enemy ground positions. The reliability of the RFC machines had exercised my father rather more than the response of the enemy and he had not ceased to regard aviation as an extremely dangerous occupation.

The outbreak of war in 1939 as I was concluding my education, the excitement of the air battles over England in the summer of 1940 and the nightly bombing of London where I was living and working for a merchant bank made the choice of the RAF the obvious one, but I had been surprised when the aircrew selection board had positively urged me to opt for pilot training. Without a licence to drive a car or even a motor cycle I had assumed that I would be assigned to training as a navigator or wireless operator. That this did not happen was probably due to the great number of eighteen year old volunteers who were equally inexperienced.

In the spring of 1941 I reported to the RAF Receiving Wing in Babbacombe, Devon where former civilians were inoculated, lectured and supplied with uniform and flying kit. Most of us were under the age for call-up, a favoured group for aircrew training as were volunteers from overseas, but others who arrived there already in uniform had been engaged in various ground duties, having applied for aircrew selection after their call-up.

We were accommodated in seaside hotels and boarding houses which the RAF had requisitioned. I shared a room with two brothers from Uruguay called Surgey, Robin Hampshire from Brazil and Oliver Wells, son of the Member of Parliament for Bedford. Two of Oliver's brothers had already been killed in action and he was sanguine about his own chances of survival, but he did so even though his Lancaster was shot down over Germany and his parachute became entangled with the tailplane after he had baled out. Amazingly the bomber crashed into the Rhine, permitting him a reasonably soft landing. During the four years of our war service Robin and one of the Surgeys were killed in action.

Most of us suffered little lasting effect from the inoculations but one man mistakenly joined each of about six queues for the typhoid and tetanus jab, collapsed and subsequently lost much of his hair before he recovered. The lectures varied from the perils of venereal disease to exciting accounts of aerial combat. These were delivered by participants in the Battle of Britain and surviving pilots of earlier action in France when their Lysanders or Fairey Battles were outgunned and outmanoeuvred by the Luftwaffe machines. Physical training instructors told us that their object was to reshape our bodies into the better condition which nature had intended them to be, while we waited for our uniforms to be tailored to fit. We learnt the strange jargon of drill sergeants: "Let's be having you . . . get fell in . . . you play ball with me and I'll play ball with you."

Although ranking as lowly aircraftmen, second class, paid half a crown a day, of which we were honour-bound to give sixpence a day to the RAF Benevolent Fund it was impressed upon us that we were cadet pilots, probable future officers and were required to wear a distinctive white band on our forage caps. Even this liberal treatment was too harsh for a former Oxford undergraduate who moved into the best local hotel, making sure that he was called in time for the first parade. Rising early one morning he found himself competing with a Group Captain for the bathroom and was swiftly forced back into our boarding house.

The issue of flying kit was an exciting foretaste of things to come and it was a common practice for cadets to dress themselves in both inner and outer suits, the high boots, leather gauntlets and helmet, attach the goggles and have their photograph taken. Not everyone was going to pass through each stage of the course but those who failed would at any rate have some evidence of the attempt. Provided with our newly tailored uniforms we were marched the few miles to the Initial Training Wing (ITW) in Paignton, accommodated in other small hotels and treated to more lectures and instruction. Here Air Commodore Critchley was in command. A Brigadier General in the first World War, he had become President of the Greyhound Association and was also reputed to own many of the hotels requisitioned by the Royal Air Force in Devon. He treated us soon after our arrival to a rousing speech and a warning always to "watch for the Hun in the sun."

A few days later our physical training instructor was detailed to take us for a route march. Corporal Biddle was a gentle man who had been a professional racquets champion. The march itself was of rather short duration ending on a grassy bank below a large house which overlooked the ocean. Calling a halt he declared that we would spend the remaining time relaxing on the grass to enjoy the view and the late spring sunshine. Our reveries were soon interrupted by an explosion of wrath from the house behind us where, unknown to Biddle, the Air Commodore resided. Our route march was hastily resumed, the dismayed corporal having been promised disciplinary treatment.

After the long winter 'blitz' on London I enjoyed the peaceful nights and recall only one air raid warning when German bombers passed nearby bound for Plymouth and Devonport. Another reminder of violent warfare was the appearance of numbers of wounded RAF officers undergoing treatment at a local hospital, causing thoughtful reflections of the possible future fate of some of us. Those officers who chose to wear uniform must have found it wearisome continually having to return the respectful salutes of hundreds of admiring cadets.

Within eight weeks a group of us were posted to the Elementary Flying Training School at Watchfield in Berkshire. Before the war the school had been a private concern and its director, whose idiosyncracy it was to carry a swagger stick, had continued in charge as a Wing Commander. The Chief Ground Instructor Captain Meager had been navigating officer of the R100 airship. He told us to forget any navigation instruction we might already have received. We would learn the true methods from him. The former civilian mechanics still swung the propellers of the de Havilland Tiger Moths on the command "Contact" but the Chief Flying Instructor and the other instructors were all RAF men. They included former fighter pilots, some of whom were Polish, but my own instructor following my initial flight with Sergeant Glauber had been a school teacher whose aptitude had been noticed at his own EFTS. He had been put on a further course to emerge as an instructor on Tiger Moths; this was Sergeant Elliot. We cadets were also promoted to the rank of Leading Aircraftman and our pay trebled to seven shillings and sixpence.

There was no accommodation on the airfield so we were billeted on householders in Faringdon. My companion in a farmhouse which after the war became the offices of Tucker's Nurseries was John Gilbert, whom I had met on our first day at Babbacombe. Despite the meagre billeting allowance of a few shillings a week the Tuckers always put out for us a plate of cold food and glasses of milk, already a severely rationed item, a welcome snack on our return from the airfield. Subsequently we were moved away when the bedroom was required for a relative whose house had been bombed but our next host was equally kind. Mr Emmans worked for the Gas, Light and Coke Company, his home one of a row not supplied with electricity and with a narrow strip of garden, at the bottom of which sanitation was provided by a thunderbox in a shed. John and I shared a room with the manager of a newsagent's shop, a phlegmatic young man, rejected for military service on account of poor eyesight. His constant complaint was the difficulty of retaining young female staff, wed or unwed, due to their propensity for early pregnancy.

Each morning the cadets emerged from their various billets to board the bus sent out from Watchfield and when we were not flying there were classes on navigation, meteorology, maps, aircraft recognition and so forth. The recognition of British and enemy aircraft from silhouettes of machines in head-on and other attitudes was a skill best learnt by small boys, but after a while most of us could pick out the different features of aircraft with in-line or radial engines, distinctive tailplanes and the arrangement of cockpit windows. Swift recognition could save the lives of a whole crew and a pupil who was heard to remark that he couldn't tell a Hurricane from a Messerschmitt 109 earned the instant rebuke "Famous last words!" An armaments officer explained the workings of the Vickers gas-operated gun. It was unlikely that any RAF aircraft were still equipped with these venerable weapons except perhaps in distant Aden or India's North West Frontier but at a time when the Home Guard was looking forward to the replacement of its pikes, the Vickers guns were probably the only ones available to elementary flying schools. The most memorable piece of instruction was the facetious remark that "the bullet nips smartly down the barrel hotly pursued by the gases."

It was fortunate for some of us that the CFI habitually passed the great majority of cadets through to the Service Flying Training Schools (SFTS). Some CFIs failed up to three quarters of each course. At Watchfield slow learners were kept back when it was thought that with further training they could prove proficient pilots. A few of these joined our course. There was also a cadet who had passed his tests and to express his exuberance and joy had taken the opportunity during his final solo flight to 'beat up' Faringdon in a series of dives and other manoeuvres. The local constabulary had found it easy to read the aircraft's identifying letters on the wings, informed our commanding officer, and another flying career was interrupted by the inevitable courts-martial. The RAF lost altogether too many aircraft engaged in unauthorised low flying by inexperienced pilots to ignore this practice.

We began our flying instruction and within a couple of days 'circuits and bumps' were combined with aerobatics such as looping the loop and rolls off the top of the loop. The object of aerobatics was to teach the pupil to retain control of his aircraft in whatever attitude it might be, inverted or in a steep bank or climb. Additionally the instructor would put the machine into a steep climb, close the throttle to induce a stall, kick one rudder pedal forward so that at the moment the aircraft began to dive it also entered a spin, the fields below whirling round in a dizzying manner. Then he would demonstrate that the only way to recover control was to centralise the rudder pedals and lift the nose by moving the control stick, not backwards – the instinctive reaction of the novice – but forward. One could not watch his hands making these movements as one might watch a golf or tennis coach because only the back of his head was visible from the rear cockpit and he communicated through a tube, but the pupil's hand was on the control stick and his feet were on the pedals, so movement was felt.

I had rather dreaded the prospect of aerobatics, fearing that I might be sick. The thought of flying inverted in an open cockpit, held to one's seat by a couple of straps, was singularly unappealing. Before volunteering I had made discreet enquiries to discover whether parachuting practice was involved. I was quite prepared to jump out of an aircraft to save my life but the idea of projecting myself into space in cold blood filled me with vertigo. My fears were allayed on these scores and I found myself enjoying looping the loop and then completing a roll off the top. I got no pleasure from flying inverted for any length of time, but aerobatics practice did give me confidence that at all times I could master the machine.

Sergeant Elliot also showed me how to restart the engine in flight if it cut out by putting the aircraft into a dive, and in another exercise he warned me that on some future occasion he would close the throttle to simulate engine failure and expect me to glide down to a safe landing on one particular field which he pointed out. The whole countryside was covered with fields which to me looked no different to the one he had indicated but this was used for instruction because the ground was sufficiently flat and firm for a Tiger Moth to land without damage. On the two subsequent occasions when he abruptly closed the throttle and told me to land I looked wildly around, failed to identify the correct field and began a glide to the wrong one. Exasperated, my instructor opened the throttle to climb out of trouble but acknowledged that the exercise was really somewhat impractical.

Most of the cadets were sent up on their first solo after about ten hours dual flying and those whose instructors remained unwilling to permit this grew fearful that they might be posted away as failures. One Irishman had already left but we knew that this was because his chosen method of landing was to close the throttle at whatever height his aircraft might be as he crossed the boundary and aim it almost vertically at the ground. Pupils had been told that it would be no disgrace to acknowledge at this early stage that they had made a mistake and did not wish to continue to train as pilots. One of my colleagues told me in strict confidence that he was frightened the whole time he was in the air.

"If you asked to retrain as a navigator it would be worse," I told him. "Your life would then depend upon someone else."

"I know that," he admitted, "and I would probably be just as frightened in a submarine or in the front line with the poor bloody infantry. I am going to try and stick it out."

I was very sorry for him because by that time I was enjoying myself and satisfied with my own choice.

The Tiger Moths were not equipped with radio telephony. The pupils were taught to note the direction of the wind from the windsock or the large letter 'T' which lay on the ground in front of the control tower, and keep a wary eye upon other aircraft in the circuit as well as those landing or taking off. Pupil pilots concentrating almost exclusively upon their own approach or take-off and seemingly unaware of the close proximity of others caused great apprehension to their instructors watching from the ground.

I had just completed ten hours dual and John Gilbert had already gone solo when the weather changed for the worse and the CFI decreed that no pupil should be sent on a first solo until there was an improvement in the conditions. This was a disappointment as pupils took great pride in being among the first to go solo and the extra dual hours in one's log book might indicate slow progress. When the gusty winds abated and I was allowed to take off on my own I was immensely relieved and with all the unjustified confidence of youth I was delighted with myself when I completed the circuit of the airfield and made a respectable landing.

The next exercise was cross-country flying to destinations an hour or more from Watchfield. I made all the usual mistakes, trying to identify on my map places I could see on the ground instead of working out from my course and the time and distance flown what I ought to see from reference to the map. I had great difficulty turning the aircraft on to a course demanded by my instructor. The Tiger Moth did not possess a gyro, the compass needle swirled about in the liquid bowl for some time after I had levelled the wings and then I would find that I would have to make a further alteration of thirty degrees or more. Occasionally I forgot to tighten the friction nut on the throttle so that it began to close as I removed my hand to alter the compass grid.

When Sergeant Elliot was satisfied that I could be trusted to fly further afield I embarked on a cross-country to Luton. I succeeded in finding the grass field there, touched down smoothly and reported my arrival in the control tower, presenting a chit on which the duty officer was required to write the time and quality of the landing.

"I was pouring myself a cup of tea and didn't notice your touchdown," he remarked. "What sort of landing was it?"

"Oh! excellent," I claimed, "one of my best."

He gave me a doubting glance and wrote "satisfactory."

One cadet on my course suffered a genuine engine failure on his cross-country exercise. Noting the direction of the wind from the smoke issuing from a factory chimney he selected a field which he hoped had firm ground and glided down for a safe landing. Our CFI was pleased to hear his report over the telephone but puzzled when he flew over with a mechanic to rectify the trouble. The cadet had chosen quite a small field alongside an extremely large aerodrome.

On my next cross-country I reported to the control tower as another pupil was emerging, one of my contemporaries at Paignton who had been posted to a different EFTS. He was in a sombre mood.

"Our CFI has washed out all but about seven of us," he told me; "I have my own final check with him next week and if I don't satisfy him then that's my lot."

My own final check was also due in the immediate future and this conversation made me reflect upon aspects of my own flying which were far from polished. On the plus side I did not have much difficulty making acceptable landings. This had proved the undoing of many pupils who could ease the aircraft gently into the air, perform all the necessary manoeuvres perfectly well but whose touchdowns were reminiscent of the boundings of an energetic kangaroo. Halfway through the course I had flown with the CFI in a progress test which had appeared to satisfy him but the important one lay ahead.

In the event I came near to making a serious blunder. After twenty five minutes or so of steep turns and aerobatics he told me to land and through carelessness I found myself distinctly high on the approach. I was fairly sure that I still had enough room to land and stop in the distance which I had left myself but thought it imprudent to try.

"I am going round again," I announced into the tube.

"I should bloody well hope so," came the blunt reply.

My second approach and the landing gave me no cause for concern but the CFI said nothing to me after I had taxied in and switched off the engine. He walked over towards my instructor who had watched the conclusion of our flight and awaited the CFI's report. I saw Sergeant Elliot's worried expression change to one of relief and then he smiled as my judge concluded his comments and went away.

"Good grief, Jackson!" he exploded. "Why did you have to make your worst-ever approach on an occasion like this? You really are a clot. I was praying that you would have the sense to go round again. Well," a broad grin crossed his face, "you passed the test."

I had enjoyed the five weeks at Watchfield. There had been no time for any social life outside our service surroundings. When we could afford it John Gilbert and I had enjoyed a few beers and an evening meal in Faringdon's Bell Inn. There we usually found a nunber of officer cadets who were training in Shrivenham. They used to complain that the constant noise of the Tiger Moths circling their barracks interrupted their studies. That evening we had something to celebrate and we had also received word that we were to be sent almost immediately by train to Scotland to embark on a troopship for Canada.

CHAPTER 2

CANADA AND THE OXBOX

The Belgian ship *"Leopoldville"* gave many of us our first opportunity to sleep in a hammock when we sailed out of Greenock on the initial stage of our journey. Two destroyers accompanied the convoy which drifted almost motionless in the lee of the Outer Hebrides to enable a doctor to perform an operation in sea conditions which were sufficiently calm. It took three days to reach Reykjavik where we were driven in open trucks to a transit camp. Iceland had been occupied by British forces to forestall a German invasion and thereby to protect the northern sea lanes along which convoys carried supplies to the British Isles and to Russia. Although still neutral the United States Government had approved this action and President Roosevelt had sent a contingent of marines to release British troops for more active duties elsewhere. We met the marines and when we expressed surprise at the number of their medal ribbons were amused to be told the peaceful activities for which they were awarded. We remained only a few days in Iceland but had an opportunity to bathe in the island's hot springs.

We embarked on the steamship *"California"* to travel in comfortable accommodation and were supplied with excellent meals. Among us were cadets destined to continue their training at Pensacola and other military airfields in the United States. They had been issued with blazers and grey flannel trousers as the uniform of a belligerent power would have compromised the US Neutrality Act. The vessel was the largest ship in the convoy and its principal source of protection. Huge guns had been mounted on its decks but the date of manufacture clearly embossed upon them, which was 1898, was rather less reassuring than their appearance. Some months later the *"California"* was torpedoed and sunk but our journey to Halifax was without exciting incident, the only other vessel sighted being a four-masted schooner.

When we landed at Halifax I was in a group which set off on a five day journey by rail to Alberta to form the first course of pupil pilots on an airfield called Penhold. On several occasions when the train stopped we were marched around some small prairie town for the purpose of exercise, but none of those Canadians to whom we spoke had ever heard of Penhold. In the event this was not surprising because the airfield with its single runway was situated at the mid-point between Calgary and Edmonton, towns two hundred miles apart. It had served as a weather reporting station and emergency landing ground. Hutted accommodation had been added to house the RAF contingent but at the time of our arrival there were no aircraft to fly, nor had any arrangements been made to pay us, so only a very few far-sighted individuals were able to enjoy the very limited facilities for recreation. Red Deer, a small township, was ten miles from Penhold and boasted one small hotel, the Buffalo, complete with a wooden rail for the benefit of the cowboys' horses. There was one street with shops and one cinema. The telephone exchange was manned by a woman to whom one had to give the required number. They were acquainted with all Red Deer's families and one evening when there was no reply to a call I made to a girl whom I had met the telephonist remarked: "I guess Marie is washing her hair tonight."

Our first impression of our new surroundings were somewhat bleak, particularly when the commanding officer thought fit to reduce the generous rations initially provided by the Canadian caterers to those in effect in Britain. Combined with a shortage of RAF cooks the standard of food was such that orderly officers asking for "any complaints" heard quite a number, and ample quantities of butter, cheese and jam were sensibly restored to the mess table.

We were addressed by the Station Medical Officer, who warned us that we were living in an area of close-knit local communities who greatly prized their reputation for clean living. Any undesired pregnancies among their young women if attributed to the RAF presence would severely damage the good relations which everyone wished to foster.

After the compulsory church parade of our first Sunday at Penhold John Gilbert and I hitchhiked to Red Deer, walked the near-deserted streets and passed the church just as its doors opened and the girls of the choir emerged ahead of the congregation. We had paused to examine the quality of these young women when we were approached by a middle-aged man who asked us both to meet his family and to accompany them home for lunch. His daughters, who were members of the choir, did not dispel his assumption that we had both attended their church service and the kindness and hospitality of this family was to make our stay at Penhold a most pleasurable one.

The older daughter's original boyfriend whom I met one evening was a tall tongue-tied member of the "Mounties," the North West Mounted Police, reputed always to "get their man." I never learned if he finally won the hand of this girl but he certainly had plenty of competition from the RAF cadets and even from a

Squadron Leader who enjoyed the advantage of his own car. Red Deer was enlivened by the increase in the young male population and before long a band was formed and Saturday night dances proved to be a successful attraction.

Until some aircraft were delivered to Penhold we had to concentrate on classes in navigation, meteorology, gunnery, and so forth. Once more we were told to forget anything we had been taught prior to our arrival. This time the Browning gun replaced the Vickers as the object of our scrutiny and we were required to dismantle and reassemble it, with particular attention to the myriad parts of the breech block. We were introduced to the intricacies of the bomb sight whilst instrument flying practice was initiated under the hood of the Link trainer. This was kept working around the clock to the chagrin of instructors and pupils who were detailed for a session in the middle of the night. Fortunately the cadets were accommodated only three to a room so the disturbance to sleep was not continuous.

In September 1941 the Airspeed Oxfords on which we were to resume our training were flown from the east coast to Penhold by our instructors. Unlike Watchfield's grass field which enabled a pilot to take off and land into wind Penhold had a single runway and a paved taxiway parallel to it. Dubbed the "Oxbox" these examples of this twin-engined aircraft with their battered engine cowlings were not new. It proved difficult to trim them to fly "hands off" so constant attention was necessary to keep them flying straight and level. A few moments spent looking at a map could find one banking away from the intended course.

I was unfortunate in regard to the instructor to whom I was initially assigned. He was irascible, impatient and nervous, far more nervous than I was. He did all the taxying himself, "to avoid wasting time" as he put it. On every approach and landing he put his own hand over mine on the throttles with a vice-like grip, frequently overriding any movement I had made to the power, and his feet on the rudder pedals competed with the pressure I was exerting to keep the aircraft in the right direction. Not surprisingly some landings ended in humiliating ground loops accompanied by angry curses on his part.

"By God, Jackson, you will kill us both, you really will!"

I was on the point of seeking out my flight commander, ready to risk his displeasure, to ask for a change of instructor when the latter was taken off flying duties and another officer was put in charge of me.

"How do you feel about going solo?" he asked, and I assured him that I was keen to do so. I had to explain that I had not been given the opportunity to taxy the "Oxbox" and when he had got over his surprise he proposed a solution.

"I will take you out to the runway and get out. Do a couple of circuits to allow me time to reach the other end and you can pick me up there when you have landed." He scratched his head. "I have heard the expression 'learning to run before you can walk.' Now I have actually found a case of it. Well, good luck."

I took off, happy to join the majority of the course who had already flown solo in the Oxford, and thereafter my instruction continued in a thoroughly satisfactory way.

The autumn of 1941 was fine with clear skies, an Indian Summer the Canadians called it. This made cross-country navigation simple. It was always possible to reduce height and thereby read the names of locations inscribed in huge letters on the grain elevators which dotted the prairies. Some pupils flew low along the railway tracks and noted the station names on the platform. Night flying practice began and with it the need for even closer attention to the blind flying panel with its artificial horizon and indicators showing turn and bank, climb and descent. We learnt how to tell from the appearance of the twin rows of runway lights whether we were too high or too low and how much power was needed to maintain a correct rate of descent until the rows of lights almost merged and the moment came to close the throttle, raise the aircraft's nose and hopefully make a gentle touchdown.

We were not subjected to a regime of spit and polish at Penhold. As long as we worked hard and kept our bedrooms tidy there was a relaxed atmosphere but discipline was strict in regard to flight safety. The Oxfords were not equipped with radio telephony and vigilance was always essential in the airfield circuit. A number of us were practising take-offs and landings one night and an officer with an Aldis lamp was positioned near the take-off point. A green light was the signal to open the throttles and speed on one's way. I had completed several circuits when it seemed to me that I was being detained a very long time for the expected green light. I could not see the navigation lights of any aircraft on the approach and assumed that I had missed the green signal when I was performing my pre-take-off check. Delaying no further I taxied on to the runway and opened the throttles. When I had landed and taxied back to the take-off point the door of my aircraft was opened and an indignant voice castigated me for my premature action.

"Do anything like that again, Jackson, and I'll have you up in front of the station commander."

In a calmer voice he explained that he had been trying to maintain better separation between aircraft. This was Flight

Lieutenant Foxley Norris who ultimately achieved the rank of Air Chief Marshal and a Knighthood.

The winter weather set in and there were days when the temperature was so low that it was impossible to start aircraft engines. When heavy snow showers reduced visibility to a few hundred yards cadets on cross-country flights occasionally got lost. When a few weeks remained before the completion of our course my father wrote to me from Chile where my family lived and suggested that I apply for leave to visit them before sailing back to England. My application was rejected on the grounds that Chile was a neutral country and as I had been born there it might make difficulties over my return to Canada. When my parents were told this they decided to fly north to Alberta.

Their journey took five days and I met them at Calgary Airport when they stepped off a Lodestar of Trans Canada Airways. We travelled by train to Red Deer where I had booked them a room at the Buffalo Hotel. There was little for them to do during the daytime but my Canadian friends looked after them most hospitably. One morning I was startled to be told to report to the Station Commander and entering his office was even more surprised to find him chatting with my father. It transpired that an officer had met my parents in the hotel and invited my father to visit Penhold. My instructor was keen to take him up with us on a routine training flight but when the Station Commander got wind of this he vetoed it and substituted a ride in the Link trainer.

A few days before my parents were due to fly home to Chile the Japanese bombed Pearl Harbour and the United States became embroiled in the war. The American domestic airlines were required to give priority to officials on government business and there was sone confusion before my parents learned that their travel arrangements were confirmed. On the evening before their departure we went to see *"Target for Tonight,"* a documentary film depicting the crew of a Wellington bomber on a raid over Germany and their return to a fog-shrouded aerodrome with their wireless operator wounded by anti-aircraft fire. The film was a classic of its time. No actors were employed: the pilot, Wing Commander Pickard, was later to carry out a daring operation, successfully breaching the walls of a Gestapo prison through which many prisoners escaped. Subsequently he was killed in action. The young navigator, then a sergeant, survived the war to join the same airline as myself in 1947. Along with the audience the film left a deep impression on us and my parents' faces reflected their concern. This was not relieved by the approach to me as we left the cinema of a cadet whom I had known in England. He reeled off the names of our former colleagues who had been killed in accidents since arriving in Canada.

In the last week of our course we had examinations on our ground studies and Foxley Norris gave me my final flight check. We were invited to express a preference for posting to fighters or bombers. Almost every cadet had passed but it was clear that in some cases the pupil had been given the benefit of the doubt. I was surprised to see the name of one cadet who had been considered a marginal case on a list of those recommended for further training as a fighter pilot. Almost all of us had wanted to be fighter pilots.

My instructor provided a most revealing explanation. "It wouldn't be fair to his crew to put that chap in charge of a bomber."

On December 18th 1941 we were presented with our wings at a ceremony to which we were allowed to invite our Canadian friends. My new rank as from that date was Pilot Officer and my log book recorded a total of 120 hours flying, about half of it solo. The Station Commander and our instructors were our guests at a dinner held in Red Deer. It began in formal style with complimentary speeches, but following the departure of the CO and senior officers the fun became fast and furious and one young officer performed his party piece, leaping from a pyramid of chairs. This was Peter Horsley who retired from the RAF in 1975 as an Air Marshal.

The news from the war zone was all bad during our remaining tine in Canada. In the Far East *"The Prince of Wales"* and the *"Repulse"* were sunk by the Japanese. Hong Kong fell, Singapore was threatened and in Egypt General Rommel had advanced to within sixty miles of Alexandria. Travelling east in the comfort of the first class carriage of the Canadian Pacific Railway I met the same train manager who had accompanied us as cadets to Alberta several months earlier. He invited me to his home in Halifax and told me how impressed he was by the good behaviour of the RAF servicemen as compared with the army recruits bound for Europe.

New Year's Eve celebrations filled the major hotels in Red Deer as every unattached local girl was snapped up as a partner by servicemen awaiting a boat for Britain. In a province officially 'dry' taxi drivers and others had no difficulty supplying revellers with all the drink they wanted. But as 1941 came to an end every radio broadcast carried more news of disasters in the Far East and the newly trained Australian pilots were unhappy that they were about to embark for Britain when Japanese troops were successfully sweeping through Malaya and the Dutch East Indies towards their own shores.

The *"Beaverbrae"* on which I travelled to England had served as a cargo boat and accommodation was limited to about sixty passengers. These were exclusively newly-commissioned pilots of the RAF and RCAF. During the war most such ships carried one senior officer of Squadron Leader or Major rank to be responsible to the ship's captain for the servicemen on board. Our minder was a middle-aged RAF officer who wore wings and claimed to have fought in the Spanish Civil War. A few of us who had occasion to speak to him in his cabin and had noted the numerous cartons of cigarettes and liquor were made aware that he was running a profitable trade in smuggling, confident in the knowledge that the British Customs Officers paid little attention to servicemen arriving off troop ships.

We were ordered to carry lifejackets at all times when on deck and to sleep clothed as the danger from interception by a U-boat and a torpedo attack was high. We also heard that the ship's cargo was composed of explosives so the impact of a torpedo would almost certainly bring a swift death. Consequently those of us in my cabin preferred to sleep in our pyjamas. It soon became so cold on deck that we preferred to remain under cover. The pilots were rostered to take turns with the officer of the watch to assist the look-out and my turn came to perform this duty.

"Enjoying the trip?" the Mate asked me.

"Not too bad," I replied. "First time I've been on deck for three days; its been so cold. Listening to the depth charges being set off was rather worrying earlier on but its been nice and quiet since then."

"You know why, of course?" He took my blank look as incomprehension and went on: "We have used all our depth charges. Have a look round the convoy. Five ships which began the journey with us are no longer there."

I could think of nothing useful to say to that and remained silent. The journey to Britain took ten days and we anchored off Liverpool on a raw January evening, ready to enter dock and disembark the next morning. That night we heard gunfire and went on deck. Liverpool was enduring one of a long series of night raids and the blackout was breached by the light of fires from incendiary bombs and flares. When the anti-aircraft fire was interrupted we could hear the drone of the bombers. It was a sight reminiscent of the winter of 1940 in London but a new revelation to the Canadian pilots who stood watching among us.

At dawn the next day we left the ship and made our way through the ruins of the dockside buildings to board a train for the South of England.

CHAPTER 3

THE EMPIRE FLYING BOATS

Our train journey ended in Bournemouth where we were accommodated in various large hotels requisitioned by the RAF for pilots and navigators whose training had been at schools overseas. Some men had returned from the United States, an experience most had enjoyed, but they reported that the failure rate had been so high that many of the rejected cadets had been given another chance to pass the course in Canada. The successful ones had been awarded United States Air Force Wings.

The first priority for the newly-commissioned airmen was to order officers' uniforms, and many well-known tailors who had opened branches in the town were kept very busy. Bournemouth's population was also swollen by the office girls of businesses evacuated from London. The dance hall on the sea front was packed to capacity every evening and young secretaries and typists had no difficulty in securing a partner.

After a few weeks I was posted to Brize Norton in Oxfordshire. This was the first and last RAF station where I enjoyed the luxury of a WAAF batwoman to bring me a cup of tea in the morning and to polish the brass buttons of my tunic. It had been a SFTS like Penhold with the difference that in addition to Airspeed Oxfords there were also single-engined Harvard trainers. The school had been reclassified an Advanced Flying Unit, the object of which was to accustom those who had flown in mainly fine weather abroad to the more fickle nature of Britain's climate. Such pilots quickly found that it was usually impossible to navigate solely by reference to a map with the expectation of identifying specific landmarks in the course of a flight.

Pilots frequently did get lost and one day when I was completing a cross-country exercise and rolling along a runway I realised that I had not just landed at Brize Norton: it was Upper Heyford, no great distance away. I confirmed this by a glance at a notice board on my way to report to the duty officer in the control tower. I pretended that I had landed because a line squall was passing through Brize Norton and I had not wished to use my remaining fuel waiting for an improvement. As it was raining heavily at the time this explanation was accepted.

Other pilots had more humiliating experiences. Formation flying, usually of three machines, was included in our training and pilots engaged upon this practice had to direct all their attention to the hand signals of the formation leader while keeping their own aircraft close in and slightly behind him as he climbed and made turns. At the conclusion of the exercise the formation leader had to return to the airfield for the aircraft to break away and land separately. When the leader lost himself, as did happen, both the other pilots were undoubtedly lost also and all three had to put down at a strange airfield to find out where they were.

While we waited to be posted to an Operational Training Unit and thereafter to a squadron we were taught to use an approach and landing aid called the Standard Beam Approach. Many British airfields were equipped with this system. Briefly the pilot listened to a series of dots or dashes depending upon whether he was flying to one side or other of the beam. Intermittently the code letters of the particular system were transmitted to identify the airfield. The beam became more narrow as one approached the airfield, the aircraft passing over an outer marker beacon which flashed a light on the blind flying panel and another beacon just short of the runway. Described like that it might sound a simple feat, but given variable winds and turbulence it could be extremely difficult and when we were not trying to improve our performance in the air we were detailed to practise flying the beam on the Link trainer.

Filling in time like this before being posted to more advanced aircraft was rather boring. I flew to Watchfield one day and found my first instructor Sergeant Elliot still there. He greeted me warmly enough but noting my commissioned rank complained that his own application had been rejected on the grounds of inexperience.

"I have logged 800 hours," he protested, "and youngsters of nineteen with 120 hours are leaving SFTS as officers. It's not right."

Another pilot with the same surname as myself thought it would be amusing to fly to the south coast resort where he had worked as a junior clerk and make a series of low passes over the town hall. Within a matter of hours he was grounded pending a court-martial. I became somewhat exasperated during my remaining time at Brize Norton whenever I had to give my name to anyone.

"Are you the stupid clown who beat up the town hall at Brighton?" was the more polite response elicited.

When volunteers for posting to flying boats were invited I applied and was given a rail warrant and told to report to the Grand Spa Hotel at Bristol. Along with about eight other pilots we found that we had been seconded to the British Overseas Airways Corporation to act as co-pilots for a period of a year. Later this was extended to eighteen months. If we were surprised at this turn of events we were not disappointed. Many of our contemporaries had been posted to navigation schools to fly trainees around the country in Avro Ansons. John Gilbert was one of these, although subsequently he flew two tours of operations with Bomber Command. Some pilots were sent to tow drogues which were used for target practice and others towed gliders in preparation for the invasion of Europe.

BOAC's route structure had been decimated by the entry of Italy into the war, the fall of France, the threat from German troops to the British position in Egypt and the occupation by the Japanese of most of the British, Dutch and French colonies in the Far East. The landplane base at Croydon in an area being bombed almost every night was abandoned in favour of Whitchurch, near Bristol. When the attempt by BOAC to supply Malta had to be given up through the island's lack of fuel for returning aircraft the airline's pilots were usefully employed flying converted Liberator bombers to and from Canada. The prime object was to carry the ferry crews who had flown the bombers which had been bought from the Americans, but stores and important passengers were also carried.

There were several types of flying boats in BOAC service in May 1942. Three Boeing 314 aircraft had recently been acquired to operate across the Atlantic. These were commanded by the airline's most senior captains who also possessed the First Class Navigator's Licence, men such as Kelly Rogers.and Tony Loraine. Those of us who formed the initial group of RAF co-pilots were assigned to fly the Empire Class flying boats *"Champion," "Cathay"* and *"Clare"* together with *"Golden Hind"* and *'Golden Horn."* BOAC had recently lost *"Golden Fleece"* when ice forced the aircraft down off Cape Finisterre. The crew had been picked up by the Germans. Soundproofing and soft furnishings had been stripped from the cabins to reduce weight and allow a greater payload to be carried.

We were based at Hythe, across the water from Southampton, and put up at the Westcliff Hall Hotel, a comfortable place with a large garden overlooking the port. There were very few guests other than ourselves and some junior naval officers. Their task was to conduct acceptance trials on the armed powerboats built close to the BOAC slipway, a few hundred yards from the hotel. The absence of guests could be attributed to the severe bombing suffered by Southampton and also the exclusion of non-residents from areas of the south coast in the months before the invasion of Europe. The raid on Dieppe with the heavy casualties incurred both among the soldiers and the naval officers manning the landing craft was brought home to us by the arrival of young widows at our hotel.

The dark blue uniforms of BOAC with the single gold braid ring of a Second Officer were issued and our flying instruction began under the kindly guidance of Captain Bailey. A launch manned by girls in navy-blue sweaters and bellbottom trousers carried crews to the moored aircraft. Although the four-engined flying boat was much larger than the Oxford it was not much faster in the take-off, cruise or landing phase, nor was it difficult to handle. When I had made half a dozen circuits and landings another Second Officer replaced me and I went down the stairway from the flight deck to the passenger cabin to experience the sensation in that area. During the take-off some spray was visible as the aircraft rose on to the forward 'step' before climbing away. As we touched down, perhaps rather heavily on that occasion, the water displaced completely covered the window alongside my seat and gave me an unpleasant shock.

On Southampton water, however strong and from whatever direction the wind blew, it was always possible to take off or land into it. If the sea was rough some time might elapse before the required speed was obtained to 'unstick' but there was never insufficient space. What did require skill was the ability to bring a flying boat on to a mooring buoy when the wind and tide exerted conflicting forces. This had to be done sufficiently slowly for the Radio Officer, whose job this was, to lift the noose from the buoy on to the bollard on the nose of the aircraft. Some of my colleagues had sailed small boats and did not find this difficult but we were soon to find out that few of the Captains would allow us to carry out this task, nor were we permitted to take off or land on passenger services.

Aware of our meagre flying experience the Captains were reluctant to risk their hard-won licences and position in the airline. Of course when the automatic pilot failed one would have to fly for hours on end and for as long as the Captain cared to spend in the lavatory or talking to the passengers. In retrospect this was understandable. Almost all of BOAC's own co-pilots had been recalled to the RAF or been promoted to a command. The airline had never engaged navigators who were not also qualified pilots. RAF men with as few months training as ourselves were made available, as were some wireless operators. Consequently the Captain found himself with a crew of willing but very fallible new recruits.

The Empire boats based at Southampton were employed on the route to Lagos in Nigeria with stops at Foynes on the Shannon river, Lisbon on the river Tagus, Bathurst and Freetown. The two-and-a-half-hour flight to Foynes was made in daylight. We took on board a sextant, air almanac, astronomical tables, astro compass, a Very pistol and cartridges, semaphore flags and an Aldis Lamp, all of which had been banished decades ago from more modern airliners. Air to ground communication was by wireless telegraphy, there being no provision for voice transmission and in war zones the Radio Officer transmitted only brief essential messages, his main function being to decode messages sent to the aircraft such as weather reports.

My first flight overseas was with Captain Jack Harrington in *"Golden Horn."* Most of the captains had been officers in the RAF in the 1920s and 1930s, while the legendary Captain O.P. Jones had served in the Royal Flying Corps during the first World War. Harrington had risen from the ranks to become a sergeant pilot before being accepted by Imperial Airways, a tribute to his general competence. He was a dapper man whose custom it was to don his cap and gloves to perform the landing.

We were flying over Dorset in brilliant sunshine when I noticed that a Hawker Hurricane was closely following our progress, the fighter's far greater speed obliging its pilot to make a complete turn and pass us in the opposite direction from time to time.

"I expect there is an air raid alert," Harrington remarked. "The fighter boys know perfectly well who we are and that we are unarmed."

When we landed at Foynes there was another flying boat on the river, a Sikorski of American Export Airlines which together with Pan American Airways terminated their services to Europe there. When we stayed overnight in Ireland the Dunraven Arms in Adare or the Glentworth Hotel in Limerick made us welcome. The severe shortage of petrol obliged the Irish to make good use of the horse and cart but there did not seem to be any lack of food. In Britain one could not buy a handkerchief or a bar of soap without having to hand over the appropriate coupon from one's personal ration book. There was supposed to be clothes rationing in the Irish Republic but the outfitters ignored the regulation.

"Coupons! What good are they? We could paper the walls with all those we have collected." The shopkeeper dismissed my inability to produce Irish coupons as of no consequence and was happy to sell me an excellent jacket of Irish tweed.

Occasionally one encountered an atmosphere of hostility: "Cigarettes are reserved for our regular Irish customers." This was expressed with considerable vehemence but for the most part we were received with great friendliness. Something that did amaze me was the large proportion of priests who formed the clientele of Limerick's pubs.

The next stage of our flight was to Lisbon. It was customary to take off from Foynes after sunset with the aim of reaching the territorial limits of Spain before daybreak. The Germans maintained diplomatic relations with both the Republic of Ireland and Portugal throughout the war so it was a simple matter for their agents to keep them informed of our aircraft movements. One flying boat conveying a couple of German consular officials who had been interned in Britain since the outbreak of war was actually met by men from the German consulate in Lisbon when the aircraft landed on the Tagus river. It was fortunate for us that airborne radar for detecting aircraft in flight was then still in its infancy. The Luftwaffe's Junkers 88s regularly flew sorties from France over the Bay of Biscay to prey upon British ships and aircraft.

After we had slipped our moorings in preparation for the flight to Lisbon I was ordered to demonstrate the wearing of life jackets and to point out the emergency exits. The passengers were almost all men. They included senior military officers who necessarily travelled in civilian clothes and others on important government business. This was the first occasion on which I had to perform the life jacket drill and evidently showed signs of nervousness. I hurried back to the flight deck just as a launch had completed the run ahead of us dropping flares along the river. The captain wasted no time in beginning his take-off run before the tide caused the flares to drift out of line.

The aircraft was in cloud when we levelled out at 6,000 feet but there was no turbulence. Harrington leant across to me and passed some instructions which he wanted conveyed to the Flight Engineer whose instrument panel and seat was several feet behind us. The latter was a sharp-featured man of few words, about forty years of age. I repeated what I had been told to say and his eyes narrowed into slits.

"Are you telling me how to do my job, laddie?" he asked indignantly.

I flinched under his withering look and as I turned to resume my seat heard him mutter something about babes and sucklings. The Captain then told me he was going to have a word with the passengers and left me to mind the automatic pilot.

He was soon back.

"I asked you to demonstrate the lifejackets, not to scare them all stiff," he complained. "Just about every man jack of that lot is wearing the damned thing. Why, for heavens sake?"

I was about to deny giving any such orders when the First Officer addressed Harrington. He was one of the few remaining men of that rank who had not been recalled to the RAF or promoted to a command and on this flight was responsible for the navigation. I thought he was about to complain about Harrington's habit of opening his side window from time to time to spit; the navigator's chart bore evidence of this unsocial trait.

But I was wrong.

"Captain, if I am going to navigate on any other system than by dead reckoning it would be helpful if we climbed above the cloud layer so that I can get a fix from the stars or descended below it. Then I might be able to work out the wind speed and direction by tracking a white cap through the drift sight."

"Kindly remember that I am in command of this aircraft," came the reply. "The flight plan is based upon the expected winds at this level and here we shall remain."

We did.

At dawn the coast line of Spain was visible and six and a half hours after take-off we landed on the Tagus. It was my first visit to Portugal. In Lisbon the hotels were crowded by the many refugees from the occupied countries of Europe who were hoping for a visa and a passage to North or South America. Both the German and Italian airlines had flights to Lisbon and it was an odd sensation to see their crew members in the hotels where BOAC crews were also accommodated.

We flew out of the Tagus river at night and switched off the navigation lights as we left Portuguese waters. The route to Bathurst in Gambia lay close to the African coast and the navigator was sometimes able to obtain a bearing from the marine beacons which flashed their identifying letters for the benefit of mariners. Inland lay the Sahara peopled only by the Bedouin. About ten years earlier the French company Aeropostale had inaugurated a mail service to Dakar in Senegal and later extended the route across the South Atlantic to Brazil, Argentina and Chile. When the pioneer airman Mermoz had been forced down by engine failure in the Sahara he remained a captive of the Moors until a ransom was paid for his release. In 1942 Dakar was under the control of the Vichy government and we gave it a wide berth. The sector Lisbon-Bathurst took about thirteen hours and when a couple of twin-engined Catalinas were employed on the West African route their speed of only ninety knots meant a flight time of fifteen hours.

On this first occasion I had not heard of the Harmattan, a wind blowing from the southern Sahara which carried sand up to 6,000 feet. Looking out to the east and hoping to catch a glimpse of the African Coast I was puzzled by the appearance of a diffuse red glow in the distance.

"What's that?" I asked: "a huge fire?"

'That is the sun rising," Harrington replied. "Something you will see very often indeed in this job."

We refuelled at Bathurst and continued the flight to Freetown where we had a chance to sleep. It was not easy to do so in the sultry humidity of the wettest colony of the British Empire. It was the sole military outpost where officers were permitted to carry umbrellas when in uniform. The final stage of the journey to Lagos began at dawn and took us along the coastline of an area of Africa permanently green from the rains of the intertropical zone. Lagos had become an important BOAC station, a junction for flights from Egypt and South Africa in both of which the airline's crews were stationed. There was a comfortable resthouse with an excellent restaurant and swimming pool. Malaria is rife in parts of Africa so we slept under mosquito nets and tried to remember to take mepacrine tablets. Mosquito boots and a pith helmet were issued to new crew members on our arrival.

On the flight northbound departure from Bathurst awaited a signal from Lisbon that weather conditions were favourable for a landing. If this meant a delay there was a chance to swim from miles of excellent beaches. The colony had no other attraction. It rained for eight months of the year. Lisbon had much more to offer. The bright lights were a refreshing change from the blacked-out towns and villages of Britain and the warm evenings enabled one to sit outside the cafes and bars and enjoy the passing scene.

During my wartime service with BOAC I met the actor Leslie Howard in the British Club in Lisbon. He had been sent to Portugal on a goodwill mission, his face being known to millions of filmgoers. The following day the airliner in which he was returning to Britain was intercepted by an enemy aircraft over the Bay of Biscay and shot down. The Douglas DC-3 *"Ibis"* belonged to KLM and was one of several which after the invasion of Holland had been flown under BOAC auspices. On two previous occasions this aircraft had been attacked by the Germans yet it departed from Lisbon in the morning for a daylight flight to England.

Our flying boats always took off at night and landed at Foynes soon after dawn. The passengers and crew were served breakfast ashore before returning to the Custom House alongside the jetty.

"Any eggs, butter, meat?" I was asked.

"No!" I replied and gestured at the illustrated notices on the walls which expressly forbade the export of these foodstuffs. "The regulations are pretty specific."

"You don't have to bother about those, not you boys."

Delivered in the soft accent of southern Ireland the expression on the Customs Officer's face clearly indicated astonishnent that I should pay any regard to official decrees.

“But it’s only six o’clock,” I reminded him. “Surely the shops aren’t open at this hour?”

“You be going along to Hogan’s and hammer at the door. He will have to open up then.”

Enchanting people the Irish.

There was no element of luxury in wartime air travel. No stewards were carried. Passengers and crew had to make do with sandwiches, an apple and mugs of soup. The aircraft were noisy, unheated and draughty. Prior to the Allied landings in North Africa the passenger windows were covered over so that convoys of troopships and landing craft should not be observed and talked about. The long sectors in such slow aircraft were as much an endurance test for the passengers as for the crew.

During my eighteen months of secondment to BOAC I flew with about ten captains and admired their professionalism and competence, their patience with crew members of very little experience. There was a good spirit of teamwork although I recall one occasion when it had unfortunate consequences. There was a bar in Lisbon called the Olimpo much favoured by airline crews by reason of the inexpensive drinks, the lively band and a cheerful group of young women who were paid by the management to dance with and make welcome the clientele. They were not allowed to leave the Olimpo until it closed in the early hours so anyone wishing to pay for further favours had a very long wait. One night a co-pilot and navigator were leaving these premises when the manager detained them to offer a raincoat.

“Your Captain left this behind,” he explained. “He slipped and fell down the stairs, knocking himself unconscious. He had a nasty cut on his head so I called an ambulance.”

The earlier departure of their Captain had been observed by the two men but they were unaware of his misadventure. The co-pilot made enquiries as to the location of the hospital and next morning called upon the patient who was sitting up in bed, his head bandaged.

“Do you feel well enough to discharge yourself?” he asked. “If this hospital tells BOAC that you are here and where they found you the worst possible interpretation will be put upon it. Far better to telephone the company from the hotel and tell them you slipped and hit your head.” Watching the Captain considering this he added: “The Olimpo bar isn’t exactly the YMCA.”

The Captain was convinced and against the protestations of the hospital staff allowed himself to be removed to the hotel. Unfortunately for him his premature departure was reported to BOAC’s office. A disciplinary enquiry took a very serious view of his conduct and he was swiftly dismissed from the airline. The story had a happy ending. Back in the RAF and commanding a Sunderland, which was the military version of the Empire Flying Boat, the pilot spotted a German U-boat being refuelled in the Atlantic ocean and sank it. He received an immediate award of the Distinguished Flying Cross.

By October 1943 when I had completed eighteen months with BOAC, the Germans had been driven out of North Africa and in Italy Mussolini had been deposed. The airline resumed flights through the Mediterranean to Egypt and points further east. My last flight was on the sole remaining Catalina which we handed over to the Australian Airline Quantas in Ceylon. The other Catalina had been destroyed when it hit a submerged object during a landing in Poole harbour. Four of the RAF pilots who had been seconded at the same time as myself had been killed in accidents. *"Clare"* caught fire off Dakar at night: there were no survivors. *"Golden Horn"* caught fire on a test flight after an engine change on the Tagus and all but one of the crew were drowned. A flying boat struck the slopes of Mount Brandon in southern Ireland on its way to Foynes. Another crashed in North Africa, a propeller breaking free and slicing into the fuselage.

BOAC were reluctant to part with their seconded men who from the company's point of view were just beginning to be useful but I was keen to return to the RAF. I had accumulated over 1,000 hours but almost all of these were as co-pilot and I wanted to fly in command. I explained this to Captain Harrington who had becone Chief Pilot at Southampton and was hoping to persuade me to stay.

"I am fed up with sitting in the right-hand seat with so few opportunities to carry out any landings." I spoke rather bluntly and finished in the same vein. "I want to justify my existence."

For a moment Harrington was at a loss what to say. Then his face broke into a wide grin. "That's your story, Jackson. You stick to it – and good luck."

We were to meet again five years later when I found myself in BOAC again with Harrington as my flight manager.

CHAPTER 4

WIMPEYS OVER ENGLAND

Aircrew seconded to BOAC were for administrative purposes on the strength of the RAF station at Filton near Bristol and I reported there to await a posting. The principal activity at this place centred on modifications to Bristol Beaufighters, twin-engined machines which were used for low level attacks on shipping and ground targets. Normally they carried a pilot and navigator. The crews remained at Filton for a couple of days and then flew north to Port Ellen on the Isle of Islay to carry out firing trials before joining their squadrons.

A few days after my arrival it happened that one of the eight Beaufighters which had recently taken off for Port Ellen had failed to arrive there, nor did it land at any other airfield. The weather had been atrocious and none of the pilots had arrived at their destination, most returning to Filton whilst a few had diverted to other stations. The Station Commander sent for me and pressed some documents into my hands.

“Here is a job for you,” he said. “In the next day or so someone will come upon the wreckage of that aircraft, a little longer perhaps if it crashed in the sea. I want you to conduct the enquiry.”

The weather cleared and as he had predicted the Beaufighter was found. It had struck high ground near Machrihanish on the Mull of Kintyre, killing both occupants. The crash site was only about twenty five miles from Port Ellen. There was a Fleet Air Arm station at Machrihanish and occupying the navigator’s seat I set off in a Beaufighter to try and find out why the accident had happened. When we were on our way I was surprised to notice that my pilot had put the aircraft into a steep dive over countryside composed of pasture land and scattered farmhouses. Then he pulled the nose up and we swept round in an arc before he made a further swoop. Looking out, I saw the object of his interest, a girl waving from outside a house. Of more concern to me was a line of electricity pylons crossing the area. My pilot contented himself with two more runs over the house and then continued the flight without further adventures to deliver me to Machrihanish.

I refused to be impressed by the quality of his airmanship and was in no mood to be grateful for the ride.

“Thank you for not incinerating us both in the power lines,” was my parting remark.

He was unrepentant. “I have been trained to carry out low level attacks on enemy targets including those in the Norwegian Fjords. That means avoiding mountains and ships’ masts. My little foray back there was child’s play.”

Walking past a number of Fairey Swordfish I found my way to the officer’s quarters where a pretty young Wren allotted me a ‘cabin.’ I found that the bed had blankets but no sheets and when I commented upon this she had a ready response.

“In the Navy officers supply their own linen.”

It came as a relief to me to discover that some other unfortunate individual had already performed the unpleasant task of identifying the bodies of the Beaufighter’s crew. It only remained for me to set off with a driver to look at the crashed aircraft. The road petered out well short of the high ground and I continued on foot. I had a map and on that fine day I had a good view of the landscape in every direction when I came upon the wreckage. It was clear to me that whatever mistakes the crew had made in timing their descent they had also been unlucky. If the aircraft had been flying only fifty feet higher it would have missed the highest point on the Mull of Kintyre. As it was the scene of destruction made it obvious that death must have been instantaneous.

It seemed most probable that the pilot had begun his descent after he had cruised for the time worked out on his flight plan. Still flying in cloud he had mistakenly assumed that the Beaufighter would be safely clear of ground and over the sea when he had descended through the base of the clouds. After their departure from Filton there had been no communication at all by the crew. This suggested to me that the navigator had crawled through the narrow space separating him from the pilot in the hope that the improved view from the cockpit windows would enable him to work out their position as soon as the visibility permitted it. These observations formed the gist of my report.

The pilot of a Bristol Blenheim gave me a lift to his airfield in Shropshire and I went into the control tower to find out if I could obtain a further ride on to Filton. The duty officer pointed to a single-engined Stinson Reliant on the tarmac and told me that its pilot, a member of the Air Transport Auxiliary, generally known as “ancient and tattered airmen,” would be continuing his flight to White Waltham when he returned from lunch in the officers’ mess. When he reappeared he was quite willing to go out of his way and as we followed the railway lines and other landmarks which were obviously familiar to him he explained that he had no need of a map so long as the cloud base was not impossibly low. This was reassuring, but somewhat alarming was the fact that he had only one hand and when he used it to adjust the throttle setting during the approach and landing he rested the stump of his other arm on the top of the control stick.

Back at Filton I completed my report on the accident to the satisfaction of the CO. But I remained puzzled why only this one pilot had been so determined to reach Port Ellen when all the others had either turned back or landed somewhere else. The adjutant offered the most likely explanation. On the day before the accident an aircraftwoman had complained that the pilot had made very unwelcome advances to her and he had been brought before the station CO. Because he was about to leave the airfield he received no more than a severe reprimand. Undoubtedly he had no wish to be seen at Filton again.

While I had been flying for BOAC a new organisation called Transport Command had been set up to undertake troop movements. That this had not happened before the war was due to the reluctance of the Air Ministry to allocate scarce resources to a body which the Army and Navy would also wish to use. I was ordered to report to their headquarters to discuss my next posting. I was attracted to the idea of ferrying aircraft across the North Atlantic but was told that there was no call for flying boat pilots. I had read about the aerial supply route from India 'over the "Hump"' to China and asked if that was an option.

"We have one squadron of Douglas Dakotas on that route," I was told. "The Americans have the major responsibility in that area. We will send you out to India for a conversion course on to Dakotas."

I disclosed my recent lack of first pilot tine and it was decided to send me on the short course at Prestwick where Oxfords were used to teach pilots to fly the Radio Range, a system similar to the Standard Beam Approach but with four beams meeting at a centre point. Prestwick was the entry point to Britain of a continuous stream of American-built aircraft for delivery to the RAF and USAF. The ferry crews included many American civilians and some BOAC men. Also present were some Russian pilots whose job it was to test aircraft before their delivery to the USSR. The overcrowded hotel which served as a mess for these flight crews presented an astonishing spectacle in the evenings when the civilian ferry men played poker for very high stakes. They were watched with particular fascination by the Russians, conspicuous in their high buttoned olive-coloured tunics with enormous yellow epaulettes. The wads of dollars which changed hands every few minutes probably represented for the Russians a year's pay.

My instructor on the Radio Range course had an equable temperament but a noticeable prejudice against the Americans. It was not caused by the usual complaint that they were "overpaid, oversexed and over here." He had flown Beaufighters in the Western Desert; when raw trigger-happy Anerican troops arrived in North Africa they had on two occasions shot his aircraft down. Flying low and fast the Beaufighter looked remarkably similar to the Junkers 88 which performed a similar role for the Germans.

From Prestwick I went to Lindley near Leicester to pick up a crew of navigator and wireless operator and to fly old Vickers Wellington Is around the British Isles before embarking for India. Known to airmen as the "Wimpey" this type of aircraft had been the mainstay of Bomber Command before the Lancaster and other four-engined machines replaced them. We were accommodated in Nissen huts, about twenty officers to a hut, which was heated by a stove. The officers' mess was also a Nissen hut and a warming fire was kept going for as long as the officers went outside to chop up sufficient logs.

The Wellington was the first aircraft which I had flown to incorporate radio telephony in addition to wireless telegraphy. The other pilots who assembled there had previously been employed in other Commands. Some had completed a tour on bombers. A number of them had been instructors at flying training schools. None of us were very happy about the serviceability of the old Wellingtons. It was disconcerting to taxy back to a line of parked aircraft after landing and discover that there was no brake pressure left to halt the aircraft.

Even so I was very glad of the opportunity to be in charge of an aircraft again and to have a navigator to prevent me from getting lost and a wireless operator to obtain for us radio bearings and weather reports. On our exercises he would reel out the aerial which trailed behind in the aircraft's slip stream. If I forgot to warn him that I was about to land the aerial would be left out and it would wrap itself round trees, telephone lines or other obstacles on the approach to the runway. This made us unpopular with the local residents. On night exercises we were advised of the two letter code which changed every twenty four hours. If a searchlight unit picked us out we transmitted the two letters in morse code by light signal to identify ourselves as a friendly aircraft. One night my wireless operator replied with the previous night's code and I was blinded by searchlight beams before he realised his mistake and rectified it.

Early in 1944 I completed about eighty hours flying in Wellingtons practising single-engine flying, landing without flaps and other exercises. The pilots who had flown the improved versions in Bomber Command appreciated the damage the aircraft could absorb from shell bursts and still keep flying, but by no stretch of the imagination was it a transport aircraft, nor did our exercises bear much relation to what we would be expected to do in India and Burma where there were no Radio Ranges nor airfields equipped with the Standard Beam Approach.

This course completed, we travelled by train to Morecambe and were lodged in the seaside hotels and boarding houses which in normal times had housed holidaymakers. After acquiring summer-weight uniform and pith helmets we moved on to Greenock to embark on the *"Strathmore."* This had been designed as a comfortable passenger liner and I enjoyed the journey, sharing a cabin with two other officers. Even in wartime ships managed to provide much better meals to passengers than those the general population could expect in a British restaurant. To our delight a large contingent of Wrens destined for naval duties in Colombo were also travelling amongst us and a number of romances blossomed, becoming increasingly feverish as the weather became warmer. The black-out had to be observed throughout the voyage and we were forbidden to open the portholes. The heat became increasingly more oppressive as the voyage continued through the Red Sea and Indian Ocean. The *"Strathmore"* sailed in convoy as far as Alexandria and I recognised among the other ships the *"Orduña"* in which I had travelled from Chile to England before the war. The only hint of danger throughout the trip had been the sighting of German reconnaissance aircraft when we had been sailing near Malta. The destroyers accompanying the convoy had promptly put out smoke screens but no attack was made upon us.

We docked in Bombay and bade farewell to our Wren friends who were continuing their journey. We promised to keep in touch and I was one of those who were fortunate enough to be able to do so.

CHAPTER 5

DAKOTAS ACROSS THE "HUMP"

Travelling by train to Rawalpindi one of the severe storms which are the prelude to the monsoon season lit up the night sky with vivid flashes but no rain appeared to accompany the dramatic scene. The centre of the storm must have been a long way off. I had been told that in the early part of the war in the East the monsoon rains had been regarded as presenting too much of a hazard for continuous flying activity to be conducted. But just as it had not been expected that the Japanese army would deploy through the Malayan rubber plantations on bicycles and appear across the causeway to Singapore in a very short space of time, so fixed attitudes to flying in the monsoon had to be altered. I was soon to know what that would be like.

On our arrival at the RAF station of Chaklala in the North West Frontier Province we had to accustom ourselves to a daily temperature of well over 100° Fahrenheit. We slept in huts and obtained small fans for cooling. We engaged bearers, personal servants, to keep us supplied with water for drinking and washing and to do our laundry. Daytime temperatures were too high for the mechanics to work on the engines or to touch metal parts of the airframes, so flying training began before sunrise and concluded when we landed for lunch, resuming after nightfall. On our free evenings the Rawalpindi Club was a short journey away in a tonga whose owner bore us along at a running pace.

The Douglas Dakotas which we were taught to fly were excellent aircraft. So many were built and so satisfactory was their performance that fifty years after the first commercial flight in 1935 hundreds of them were still in service throughout the world, particularly in Central and South America. The RAF Dakota Squadrons in the areas of India bordering on Burmese territory which was occupied by the Japanese had the task of supplying the British and Indian troops who were trying to drive out the enemy. For the most part that task involved drops by air on sites chosen by the infantrymen or by landing on reasonably flat strips of land which had been cleared of tree stumps or other obstacles. In the rainy season when pools of water and mud disfigured the surface of the landing strips the landing was hazardous. It was anticipated that there might be a need to drop parachute troops in support of the army and to tow gliders to be released at strategic points. We received training in these exercises also.

The Dakota was a delight to fly and one could land it quite comfortably from a steep approach but my instructor strongly discouraged that practice. He pointed out that many of the strips cleared by the army for use by Dakotas were extremely short and that I must learn to approach at a speed barely above the stall, touching down in the first few yards of the strip to avoid overshooting at the other end. As the aircraft were being operated at up to 5,000 lbs all-up weight above the limit permitted to civil airlines in the USA, this was going to need fine judgement. Flying instructors quite rightly are no great respecters of rank or persons.

"So don't show me any more of your dive bomber approaches," I was told. "You will float too far along some short strip and perhaps write off yourself, the aircraft and the passengers at the other end."

Unlike the airlines, RAF aircraft for the most part carried only one pilot but on the Dakota a second person is required to operate the undercarriage so my navigator was always present during the exercises. My wireless operator, Ken Dyson, a married man from Huddersfield, was receiving instruction on the duties of a jumpmaster, the person responsible for the swift departure from the aircraft of parachute troops on receiving the signal bell from the pilot. It was explained to the wireless operators that they would have the opportunity to make a jump if they wished at the end of the course and to that end would receive the same instruction as the parachute man in the gymnasium on the technique to be used as they reached the ground. Supply-dropping exercises were carried out from a height of about 500 feet from the rear exit of the Dakota, the door having previously been removed. This duty was also supervised by the wireless operators who were put in charge of other aircraftmen carried to thrust out the sacks of supplies.

The monsoon rains broke while I was at Chaklala and on a navigation exercise I experienced for the first time the extreme turbulence to be encountered when an aircraft entered cumulo-nimbus cloud. Up, up, up we soared, the climb indicator showing a rate which the engines alone were quite incapable of providing. Then we encountered a downcurrent and the discomfort to my ears was another signal of our swift descent. The buffeting made it hard to keep the wings level but my intention was to get out of the thundercloud and return to Chaklala as soon as I could. The turn took a long time as I was determined not to risk the Dakota being flipped over on its back but when I had completed it and finally emerged from the clouds I was both relieved and forewarned of the sort of flying conditions I could expect to experience in the future.

Towards the end of the course we were told that the army were sending men to Chaklala to make their first jump from an aircraft.

We were briefed where we were to take them and on this occasion we would be dropping about twenty men in a continuous stream over the selected site.

"You have your chance to prove your heroism," I told Ken Dyson. "Are you going to make a jump?"

"Not on your life!" he replied, rolling his eyes at the absurdity of my suggestion.

In the event he did prove himself rather heroic. A British army instructor marched twenty Gurkhas out to the Dakota and clambered aboard with them.

"Sergeant, I would like you to jump first in the stream," he informed the appalled Ken. "It is their first jump and they need one of us to show them the way. I will hustle them out fast behind you and drop last."

The shock left my wireless operator speechless and he found himself seated next to the open hatch alongside the Gurkhas. When he found his tongue it was too late to extricate himself from his predicament without incurring the scorn of his new companions. He made the jump and suffered no broken bones.

Quite often the trainees made their first jump singly and I had an opportunity to watch this from the cockpit door when someone else was doing the flying. Each soldier was ordered to "brace," standing on the sill of the open hatch with his arms outstretched, his hands clasping the door frame. At the command "jump!" he was supposed to project himself out. Most did but a few crumpled as if in a faint, only to be heaved out by their instructor.

"They are grateful to me for doing just that," the latter told me in a confident tone. "They couldn't face their mates again if we landed with them still on board."

The monsoon rains were in full spate when I arrived at the RAF station of Dum Dum near Calcutta to join 52 Squadron. I was lodged in a house requisitioned from an Indian merchant and shared a room with another officer. We had a washbasin and shower, each providing cold water, and depended upon our bearer for hot water for shaving. There was no glass in the window frames and the nets over our beds were absolutely essential as a stagnant pool in the grounds bred mosquitoes by the thousand. A larger house nearby served as the officers' mess.

To travel the few miles to our aircraft site two fifteen-hundredweight trucks were provided for the use of the squadron's officers and it was on one of these vehicles that I learnt to drive.

The squadron performed several roles. One of these was a daily service to Kunming in China, landing en route at Dinjan in the north eastern province of Assam before crossing the range of mountains known as the "Hump." Passengers were seldom carried, the load usually comprising stores for the British Military Mission. This was headed by General Carton de Wiart, Winston Churchill's personal emissary to Chiang Kai Shek who led the Chinese troops still resisting the Japanese. The responsibility for conducting the war against the enemy in that area was mainly borne by the Americans who not only had squadrons of Liberator bombers based in Kunming but from Dinjan and Jorhat in Assam mounted a procession of flights of supplies using Douglas Skymasters, Curtis Commandos and Dakotas. There were four aerial routes across the border, the northernmost requiring aircraft to be flown over the highest ranges, the southernmost obliging pilots to fly over areas occupied by the Japanese. However there was usually a mass of cloud over all the routes and very turbulent cloud too, so interception by Japanese fighters was not a high risk. Oxygen masks were fitted in our Dakotas which could cruise at twenty thousand feet.

The method of loading our aircraft would never have been tolerated by any airline. The philosophy was that anything that could be fitted into the hull could be carried and the largest and heaviest items were loaded first to be lashed close to the crew door, the smallest items such as crew baggage last. It was rare indeed for anything to be weighed because few squadrons were provided with scales. There must have been occasions when an engine failure during take-off caused a disaster but 52 squadron was fortunate in that respect. One of our pilots had an engine failure crossing the "Hump" and ordered his two crewmen to dump the load as quickly as possible to reduce weight. They made it back to Dinjan. On a flight when severe turbulence was being experienced Ken Dyson went aft to make sure that the cargo had not broken loose, endangering the stability of the aircraft. A severe downcurrent lifted him off his feet and he landed face first further down the cabin, cutting his head.

This was the most demanding route, one pioneered before the Western Powers had entered the war against the Japanese. American mercenaries under their leader the retired General Chennault had worked for Chiang Kai Shek for several years in an organisation called "Flying Tigers." When I joined 52 Squadron they were flying Dakotas and some of our pilots made their initial flight to China with them to gain experience of the route.

One day an army captain was dining in our mess prior to a flight in one of our aircraft to Dinjan where he was stationed. He had been a tea planter and told us that he was still living in the same large house, his current job with the military being the recruitment and employment of local labourers.

"If you should ever have a delay or an overnight stop at Dinjan," he remarked to me, "you will find my house more comfortable than the transit mess on the airfield. Just mention my name and have yourself driven over."

I remembered this when I landed there and learnt that priority for flights to China was being accorded to larger aircraft carrying greater loads. Dakotas would have to wait until the following morning. As predicted a jeep was readily available to take me to the former tea planter's house. I rang the bell and a smart white-apparelled bearer appeared, accepted the explanation for my arrival and led me to a sitting room where tea, he assured me, would soon be served. A few moments later an American Colonel walked in and listened politely as I introduced myself.

"I'm sure a bed can be found for you, probably in our host's own bedroom," the Colonel said, adding, "this is the Commanding General's headquarters you know."

I certainly did not know and began to wish myself back in the transit mess. At dinner there was still no sign of the former tea planter and I felt very scruffy seated among senior officers and a three star General. Our host did not appear until I was about to part the mosquito net and climb into the spare bed in his room. He stopped dead in his tracks as he observed me through the netting.

"Who the bloody hell are you?" he demanded thickly before collapsing on to his own bed. He was asleep before I could begin to remind him of his invitation.

At Kunming a small number of RAF men under Wing Commander Lord de Waleran made sure that our Dakotas were properly serviced in time for our return flight in the morning. They included a medical officer but there was so little call for his services that he made himself available to the families of Chinese employees. On one errand of mercy his jeep which bore an RAF roundel and the usual service markings was stolen by one of the minions of the local Chinese warlord. It was often seen, its origins barely disguised by overpainting, but diplomatic efforts failed to recover it. Fortunately a jeep could be fitted into a Dakota and a replacement was flown in from Calcutta.

Another task of the squadron was to ensure a flow of supplies to forward units of the army. When this involved a landing rather than a supply drop we returned to Calcutta with their wounded and mail. The Dakota was fitted with inward-facing bucket seats which could be let down from the sides of the hull and stretchers could be laid on the floor when cargo was not carried. One place routinely supplied was Imphal on the border with Burma which the Japanese had been able to surround. The terrain there obliged pilots to take off and land on a collision course with each other so vigilance was essential. The lack of any radio beacons anywhere except at established aerodromes meant that flying into airstrips located in valleys in northern Burma, particularly during the rains, required a keen awareness of high ground in the vicinity. Pilots who descended through the cloud base to find out where they were took a great risk. General Wingate, who commanded the Chindits, a special force which operated behind the Japanese lines, was killed among his companions when he browbeat the crew of a Dakota to fly on a day when the valleys were shrouded in cloud and rain.

In contrast to these activities was a passenger service from Calcutta to Colombo where the airfield, long since abandoned, was called Ratmalana. These flights provided an opportunity to renew friendships with the Wrens whom I had met on the *"Strathmore."* There was a popular dinner dance place called The Silver Fawn. The girls naturally enjoyed an active social life because they were so much in demand and it was a source of great annoyance to non-commissioned ranks that most of the Wrens accepted invitations only from officers. The girls were under orders to record in writing the name and rank of anyone with whom they went out in the evening.

In Colombo I also sought out a family whom I had met in England before the war during my school holidays. Formerly a tea planter on the island, the head of the family had a desk job with the army in the town. He introduced me to another family still on their tea estate upcountry and I spent a most enjoyable leave with them. Time spent in the hills, either in China or Ceylon, was a good opportunity to recover from prickly heat, an irritating skin complaint, which most of us suffered for part of the time in the hot and humid conditions in which we lived and worked in Bengal.

The great majority of the tea planters were Britons on contract and the tea which they produced was exported to the United Kingdom to be blended with other teas before being sold under famous brand names. Naturally the planters took great pride in their own product and would claim that the packets of tea bought from shops were a poor thing in comparison. But one becomes so accustomed to just such packets that any other taste is not always pleasing and I found the highly aromatic tea served by my hosts altogether too fragrant.

The planters were pessimistic about their future. As the war in Europe came to an end and the Labour party won a landslide victory in the general election which followed the defeat of

Germany, the new government sent to Ceylon a group to arrange the island's transition to independence. In their clubs in the hills where tennis and bridge parties followed the working day the planters shook their heads and agreed that they could not recommend their sons to follow them in their own profession.

It had been customary for 52 Squadron to provide pilots for the personal aircraft of the most senior commanders in our theatre of war. General Slim of the 14th Army had his own Dakota as did General Carton de Wiart in China and for a short period I flew the aircraft of General Sir Oliver Leese, Commander in Chief, Allied Land Forces. Compared with the varied tasks of the Squadron it was not interesting work and I was pleased when it concluded with my posting as a flight commander to 267 Squadron in Burma.

The concluding months of the war in Europe had acquainted us with some astonishing developments in aviation technology. For the first time we heard of fighters powered by jet engines and the use by the Germans of pilotless winged missiles called "doodlebugs." Now the whole of the war effort could be directed against Japan whose forces in Burma were already in retreat southwards towards Rangoon.

CHAPTER 6

BURMA AND THE FAR EAST

267 Squadron had been involved in operations in support of the army and I joined them near Rangoon shortly after their move from Akyab. Their quarters were tents near the airfield known as Mingaladon. A row of thunderboxes over a deep trench humming with bluebottles and some cold water showers had to suffice for sanitation. The men had been worked very hard and for some months had been without anyone above the rank of Flight Lieutenant to command them. Not surprisingly the sudden arrival of a Wing Commander as CO and a Squadron Leader in the shape of myself was a disappointment to the officers who had felt that *they* deserved promotion rather than outsiders. One of these had been an instructor when I was a pupil at Penhold and felt particularly affronted.

Fortunately the war was in its concluding phase and there was plenty of interesting work for us to do. The atom bombs dropped on Nagasaki and Hiroshima obliged the Emperor Hirohito to sue for peace and senior Japanese officers flew into Mingaladon in a transport aircraft remarkably similar to a Dakota to meet their British counterparts. Four crewmen disembarked first, arranged the aircraft steps, lined up and smartly saluted their own senior officers as they disembarked. I was impressed by their good discipline.

Within a few days 267 Squadron received orders to carry out supply drops to camps in Thailand in which allied prisoners of war were held and thereafter to fly on to the aerodrome at Bangkok where British prisoners were being taken for release to us and repatriation. This was a very welcome task and as the weather was not bad I arrived over the camp assigned to my crew and flew round it at a few hundred feet to look for a good place to begin the drop. The obvious choice was an area which looked like a parade ground but I could see far too many of the prisoners waving excitedly from it and was considering whether it would be more prudent to make the drop of Red Cross supplies outside the walls and wire when I noticed that the parade ground was being cleared. The prisoners were easily distinguished from their guards by their near-naked condition, ragged shorts or loin cloths being all they wore. I was then able to discharge the load in a series of runs over the camp, finally flying over and waggling the wings in farewell before resuming the flight to Bangkok.

It was somewhat disconcerting to be joined on the circuit of the aerodrome by a Japanese Zero Fighter but he shied away and I landed and taxied to a line of Dakotas which were already there. As we opened the aircraft door we looked down on a group of happy but very emaciated men, ready to board for the flight to Rangoon. They had been given shirts and slacks but had no baggage and appeared to have no possessions at all. We had put on each seat a packet of K rations whose contents contained among other things very rich chocolate, but the miserable diet to which they had been subjected since capture several years earlier made consumption of this food difficult if not impossible. Some of these men had survived forced labour on the railway project which had cost many prisoners' lives. The flight back to Rangoon in the turbulent clouds over the mountains which marked the border between Thailand and Burma must have been an ordeal but at Mingaladon airfield ambulances were lined up ready to remove them to hospitals.

When I walked through the cabin of the Dakota towards the rear door something caught my eye among the discarded packets of rations. I picked up a small cotton pouch, recognisable as the 'housewife' issued to all servicemen to enable them to sew on buttons and so forth. Inside was a well-used toothbrush and the 'pips' normally worn to denote the rank of a lieutenant, the only items retained during captivity by one survivor.

There were more flights to Bangkok on subsequent days for the same purpose and on one occasion when I had to remain overnight a group of Thai businessmen insisted on taking my crew out to dinner. We enjoyed an excellent fish meal and I experienced for the first time the oriental custom of pressing hot towels on one's face in the course of the meal. A few days later the Squadron began daily flights through Penang and Kuala Lumpar to Singapore. The Japanese troops were still very much in evidence, still highly disciplined and carrying out various tasks assigned to them. Accommodation was provided for our crews on overnight stops but initially we found it necessary to take our own bedding with us. The local population was very friendly. They had been unable to obtain bread, tobacco and other items, and as Japanese occupation currency was now worthless cigarettes became a form of currency, being bartered in exchange for watches, leather goods, cloth and crepe for the soles of suede shoes. This type of footwear was very popular, being known as "brothel creepers."

In our tented quarters in Burma we encountered a new problem. Dacoits, local thieves whose manner of operation was to oil their bodies and intrude naked into the premises which they intended to rob, entered an officer's tent and took away the metal trunk under his bed. This was serious for him because his service revolver was inside it and loss of a weapon was an offence. We were obliged to arrange a roster whereby two officers jointly patrolled the perimeter of our site. I was engaged upon this duty one night, chatting to my companion as we strolled along and turning my torch this way and that to discourage unwelcome visitors. Suddenly the beam rested upon what looked like a coil of thick rope. Directing the torch along its length I was horrified as the beam illuminated the raised head of an enormous snake. I turned and fled, my companion hard on my heels.

As deputy to the officer commanding the squadron I had to undertake administrative duties. Charges for misdemeanours would be laid against aircraftmen by corporals and sergeants. Mothers would write and complain that their sons had not written home for over six months. Men asked for a private interview to obtain help in their domestic affairs. A wife had become pregnant by another man and wished to be forgiven. Could her husband be sent home at once to protect the family's good name? Usually not, although I recall one occasion when with the cooperation of a transport squadron based in England we got an airman home to marry his pregnant fiancée and then back East again to resume the tour overseas which he had only just begun.

Quite a different story was told to me by a pilot from another squadron. Hearing that I had flown with BOAC he asked me if I had ever visited Freetown in Sierra Leone. Confirming this I listened as he told me that after completing his training in Canada he had expected to return to England and be able to marry his fiancée whose wedding trousseau and 'going away' clothes he had bought in the coupon-free Dominion. To his dismay he had remained there to learn to fly Lockheed Hudsons and thereafter was posted directly to a squadron in Freetown. The wedding outfit he had posted home to await the day when the marriage could take place.

Worse was to follow. He began to receive anonymous letters from England informing him that his fiancée was often seen in the company of another man. After a time his letters to her went unanswered. One day a BOAC York made the first proving flight from England to Nigeria and landed at Freetown. He made up his mind to stow away when the aircraft passed through on its way north, and when it returned he boarded it without attracting notice and hid in the tail compartment. But when it was being taxied out to the runway the navigator went aft to this hiding place to check the master compass which was located there. The stowaway was discovered, ejected and lucky to suffer only a severe reprimand from the colony's most senior air officer who could probably remember being lovesick himself, perhaps in as dismal a place as Freetown.

"What happened over the girl?' I asked.

"My anonymous correspondent wrote to tell me that she married the other man wearing the trousseau which I had posted to her. She left for the honeymoon in the clothes which I had bought."

He fell silent and I became aware that he expected me to make some comment. Feeling that he had indeed been shamefully ill-used I did not hesitate to condemn the girl's behaviour.

"I think you had a lucky escape," I told him. "I wonder what would have happened if you had been able to marry her and then been posted overseas afterwards."

He shook his head, determined to nurse the hurt he had suffered. "I am sure it could have worked if I had got back to England."

Life became a little more bearable. We continued to live in tents but a cinema opened; concert parties arrived to entertain the troops; it became possible to obtain a reasonable dinner in the town, and a newcomer to the squadron, Emmanuel Galizine, informed me that he was acquainted with several girls of the Ballet Russe de Monte Carlo which had arrived in Rangoon. He suggested that he should accompany me to their quarters to invite the girls back to our mess for dinner.

"We can only get two of them in my jeep," I reminded him.

"What could be more delightful?" he responded.

I told the mess secretary to make some sanitary provision for two female guests and we drove into town to collect the girls. The evening was not a success. A few of the dancers did know Galizine but they were not at all keen to travel to the airfield along the appalling road which for many months had been the target of our fighters and bombers. Finally a couple of girls were persuaded but it was so late when we reached our mess that the dining tent had already emptied and raucous voices from the bar area were already singing "Goodnight ladies." After a few drinks the girls asked to be taken back to town and one of them was directed towards the little tent specially erected for their use. She returned looking distinctly puzzled.

"Anything wrong?" I asked.

"What exactly was supposed to be in there?"

"Well," I answered, "an Elsan, surely?"

She shook her head. "No, just one empty beer bottle."

As the Japanese handed over their control of the former colonies of Britain, France and the Netherlands, civil administrators arrived and had to be flown in to resume their former duties. I set off on a journey which was to culminate in Hong Kong but began with a stop in Singapore. On my aircraft among the police, magistrates, military officers and others was Frank Owen, the editor of the forces newspaper SEAC. He was keen to see some of the night life for which the port had once been famous. "The Great World" was a covered pedestrian concourse offering among other entertainment Chinese plays and dancing to Western music. I stood and watched a play for a time. It was incomprehensible to me but a fascinating spectacle, particularly the stage hands who were visible throughout the performance, sometimes sweeping the stage, at other times seated or moving the props. It reminded me that before the war a London theatre had put on a play called Lady Precious Stream which had tried to recreate an authentic Chinese production. The dance floor was available to those who bought a ticket and offered it to the taxi-dancer of their choice. I have always enjoyed dancing but I found little appeal in dancing with someone with whom one could not converse. None of the girls spoke English.

Our next stop was Sourabaya in the former Dutch East Indies, where a small number of RAF men were on hand to meet us. I heard firing and could see shell bursts uncomfortably close to the airfield. In a building being used as an operations room an officer was colouring on a map areas where armed guerillas were known to be active in opposition to our troops. I asked what was going on.

"We are not too popular with the natives here," he explained. "They don't want the Dutch back and think we have come to prepare the way for that outcome."

He revealed that the Dutch civilians interned by the Japanese were in very grave danger from Indonesian reprisals. It had been necessary to keep the Dutch in the camps and under the protection of armed Japanese until it was possible to remove them from the area.

A Dakota crew which had made a forced landing had been murdered by the local people and we were warned to be very careful. The Indonesians had adopted as their flag of Independence the red and white colours from the Dutch flags which were red, white and blue. We saw armed men in cars which had originally belonged to Dutch residents and had more recently been seized from the Japanese. They refused to accept any other currency but the Japanese Occupation money. To make shopkeepers change this attitude the British authorities were issuing the troops with considerable quantities of these notes, hoping to convince the civilian population that their continued acceptance of these would only impoverish them.

The journey was continued to Labuan Island off Borneo where Australian troops had arrived. After an overnight stop a Brigadier among my passengers was incensed to discover that a crate of Scotch whisky which he was conveying to General Gracey in Hong Kong had been stolen. He complained that I should have detailed a member of my crew to guard the aircraft. The thief was undoubtedly the Australian sentry whose job that had been.

When we arrived in Saigon, capital of the former French colony of Indo-China, the political situation was very tense because the Vietnamese also wanted immediate independence. Here too armed Japanese troops were being used to police the town until a sufficient allied force could be moved in. I was supposed to make a direct flight from Saigon to Hong Kong but was most reluctant to do this without having another airfield available for diversion in bad weather. I had been warned that I would have difficulty obtaining fuel if I landed in Canton or elsewhere in China under the control of some local warlord. It had not been possible to get so much as a sketch of the Hong Kong area and all I knew was that there was an airfield there. Even today the approach to the inland end of Kai Tak's long runway is quite exciting. In 1945 a very short runway was located on the edge of the bay with hills at either end. I decided to fly from Saigon to Manila in the Philippines and then make the trip to Hong Kong with sufficient fuel to return if I couldn't get in. Descending below the cloud base over the ocean as we approached the coast of China we found the entrance to the harbour and looked about for the airfield.

"It is over here, the north side of the bay." From the right-hand seat the navigator had noticed a few aircraft on the ground.

We looked for a windsock to indicate the wind direction and saw that it was hanging limply down, making a right-hand circuit and approach over the lower of the hills the safest choice. We were accommodated in a villa near the airfield and overlooking the bay on a road which today is lined with vast high-rise buildings. I noticed that the drawers of the cupboard in my bedroom were lined with an English language newspaper and taking the pages out I found that the date of publication was during the Japanese occupation. The news items where they related to the war owed their origins to military directives but I was amused to observe that the advertisements inserted by restaurants offering European fare were carefully phrased to exclude any reference to French or other cuisine which might offend the occupying power.

My Dakota was due for some routine maintenance and as I required the assistance of only one person to raise and lower the undercarriage I gave my crew the day off. When I arrived at Kai Tak the aircraft was ready and I agreed to the request of some of the aircraftmen who had done the servicing to come on board for the ride. As I was going to have to make a right-hand circuit I decided to fly from the right-hand seat and from this unaccustomed position I started and ran up the engines, made a perfunctory check of the instruments and turned onto the runway in readiness for the take-off. The aircraft carried no load but the runway was short so I opened the engines to full power on the brakes, released them and the Dakota rapidly gathered speed. As it did so I found increasing difficulty in keeping it straight and hastily closed the throttles and reapplied the brakes to stop before the end of the runway. Then I found the cause of the trouble. Someone had wound the rudder trim as far as it would go. It was my responsibility to have checked that it was in the central position before take-off. I taxied back to the take-off point and resumed the air test without further incident. When I left the cockpit and walked through the cabin to the exit one aircraftman remained seated with his head in his hands.

"I am very sorry, sir," he said. "As soon as I heard the throttles slammed closed I realised what I had done. After winding the rudder to full trim I went outside to make a visual check of the tab. Just then the truck came to collect us for lunch. I had meant to centralise the trim when I came back but I forgot."

On my return to Rangoon I found that an atmosphere of discontent was spreading among the ground crews. The first reports about the plans for demobilisation had given the impression that men who had already spent several years overseas could expect to remain for many more months. The atmosphere was much worse in Calcutta. Some RAF station commanders had instituted parade ground inspections and drill sessions to occupy men without much useful work to do. Something very akin to mutiny prevailed.

267 Squadron fortunately continued to perform an important role so was not seriously affected, but I was not sorry when I read an Air Ministry Order entitling home leave to men who had volunteered from neutral countries. This followed the rapid departure from among us of Dominion and Commonwealth airmen for repatriation and release.

For some time I had been feeling ill; I had sores in my mouth which nothing would cure and I had lost a lot of weight. At first the squadron commander would not let me go until another officer was appointed to take my place. When the squadron medical officer told him I had early tropical neurasthenia he relented and made amends by arranging for me to fly home in a Sunderland flying boat. This spared me a slow journey through transit camps by train and boat.

Early on a cold morning in January 1946 I looked down upon the familiar shores of the Isle of Wight and some minutes later the aircraft made its landing in Poole harbour.

CHAPTER 7

THE CHILEAN AIRLINE

When I had joined the RAF the merchant bank who had employed me undertook to subsidise my service pay by an annual amount of forty five pounds. On the eve of my departure for India I had called in at their head office in London and mentioned that if I found it possible to obtain a job as a civil pilot after the war that might well prove to be my chosen option. Shortly thereafter I received a letter terminating the annual payment.

It remained my ambition to continue flying and whilst in Burma I had written to a newly-formed company, British Latin American Airlines, offering my services after demobilisation. I had received a reply informing me that I should call on them when released and that I would need to acquire both the civil pilot's 'B' licence and a navigation licence. With weeks or even months to wait for a boat to South America and without a home or close relatives in Britain I had a perfect opportunity to devote myself wholeheartedly to the acquisition of those licences.

About the only place where tuition could be obtained was the School of Navigation associated with the University of Southampton. I enrolled for their course and found lodgings in a comfortable house in the town, travelling the short distance to the school by tram. Among the other students were a few pilots who had already been demobilised, others who had been seconded to BOAC and hoped to receive contracts from that organisation when they had gained the licences and several young women who had been flying for the Air Transport Auxiliary.

The 'B' licence presented less of a problem than the navigation licence which involved examinations in magnetism and compasses, radio and radar, meteorology, maps and charts, air legislation and so forth. At that time there was an examination in tides to cover the possibility of a pilot being employed on flying boats. Therefore one was also required to learn the laws of the sea and the flags and lights displayed by ships. To give just one example it was necessary to recognise a trawler trawling, *under* way but not *making* way.

Much of the matter one needed to learn was of academic rather than practical use. The initial response evinced by most of us was impatience and irritation that having flown for five or more years in the most difficult conditions likely to be encountered we were expected to absorb information of a complexity that bordered upon the esoteric. That reaction had to be suppressed and there was no alternative to hard grinding study. Some of us had never even seen any airborne radar equipment, nor did the School of Navigation possess any examples. Two systems known as Gee and Loran had been developed and were available for use by civil aircraft so we had to learn the principles. The RAF and some airlines employed craftsmen specifically to adjust the various types of compasses. We had to learn how to do it ourselves. Navigation included an understanding of the use of the air almanac and astronomical tables not only to keep track of the aircraft's position by the use of a sextant but to work out where and at what time the sun would rise and set in the course of a long flight.

I spent many weeks in Southampton and almost as many weekends in Buckinghamshire at the home of very old friends to whose daughter Patricia I became engaged during that time. When I entered my name for the examinations I had no great confidence that I would be successful but hoped to pass in sufficient subjects to gain exemption from taking them again. There were so many candidates that the examinations were held in the ballroom of the Carlton Club. When they had concluded I believed that I might have passed in most of the subjects but was sure that I had failed in the radio and radar paper and probably meteorology also. This proved to be the case and I was notified that I should retake those two subjects and would be required to obtain at least seventy per cent of the available marks in each of them.

I reentered my name, paid the fees and continued to study, and the wedding was arranged for the day after the examinations had ended. I had still not been offered a passage in a boat to South America but the official demobilisation date of my age and length of service was getting so close that the Air Ministry informed me that if I wished to avail myself of a passage it would be a one-way trip. My inclination was to forego the passage and find a job but when I disclosed this to my parents their disappointment was so strongly expressed that I decided that I would have to go. With the offer of passages for both my wife and myself following almost immediately we set sail before the results of my examination had been published. Only as our ship lay outside the harbour of Buenos Aires did a cable arrive notifying me of success.

Following some correspondence I had been given to understand that I would be offered employment in a new airline to be formed by the leading Chilean shipping concern. Unfortunately the presidential election which coincided with my arrival resulted in the return of a government which was not prepared to allow any competition to the state airline Linea Aerea Nacional. The latter was the one I joined, along with several former RAF pilots who had also been born in Chile and were therefore elegible. The fact that none of them were required to pass any examinations such as those I had worked for did not seem to matter. LAN normally recruited its pilots from the Chilean Air Force. By offering pay which was marginally higher it ensured that there was no shortage of applicants but the salary was extremely low.

We did have to pass a medical examination which was conducted in a military hospital. I was startled by the amount of blood which an orderly extracted from my arm and felt like asking for it to be put back when it had been examined. I suppose that this was a Wasserman test. When I went along to the pathology department expecting to be asked to leave a sample a fierce woman in a white coat said: "Have you brought it with you?"

I mistook this remark for a disguised enquiry about my readiness to perform and said: "Yes."

"Well, where is it?" she asked.

Finally I had to see a psychiatrist. He was a pleasant man with a good command of English and we conversed on fairly general topics before he enquired whether I had any particular anxieties.

"Yes," I told him. "I don't see how I am going to be able to live on the salary LAN are going to pay me."

He thought that was very funny. "You are sane enough!" he said in dismissal.

The airline's chief instructor, Halley-Harris, was the son of Australian immigrants to Chile and bilingual. He confided to me that he was very keen to develop a modern professional outlook among the pilots but was aware that those with the longest service flew 'by the seat of their pants' and were a law unto themselves. The most senior one of them had been grounded for alcoholism. Very few of the pilots had much night-flying experience so services were only scheduled during the hours of daylight. In fact Santiago's international airport at that time, Los Cerrillos, did not possess an illuminated runway. Chile extends about 2,600 miles from its northern border with Peru to Cape Horn at its southern extremity. With the exception of a daily flight to and from Buenos Aires in Argentina all the flights were to destinations within Chile using airfields not all of which had a paved runway. A few were simply fields. Some had firm soil and one was a disused racecourse. Not all of them had a radio beacon available for an instrument approach.

LAN operated Dakotas, Lockheed Lodestars and the smaller Lockheed Electra. Two pilots and a stewardess formed the crew, a wireless operator being carried only on the Buenos Aires service. No flight plans were prepared and no route books listing radio aids and other facilities were published. Once cleared for take-off the captain turned the aircraft on to course and made his way by reference to familiar landmarks. Not too much notice was taken of the compass reading and I did not find any compass which had been corrected after 1942. Most of the aircraft had autopilots but no one used them. This was the era when the President of Eastern Airlines of America, General Rickenbacker, had refused to incorporate autopilots, maintaining that his pilots were paid to fly.

I began flying with LAN in the late spring of 1946 when the skies were usually clear of cloud. I enquired how the pilots found their way during winter when there was an overcast and was told that the Andes mountains which run the length of Chile and form the border with Argentina had a series of foothills whose peaks and ranges were often visible above the upper cloud layer. These had recognisable features. I bought an exercise book and set about producing my own personal route book based upon what I could learn from the Captains. The younger ones were more friendly and helpful than their seniors and more willing to allow co-pilots to share the take-offs and landings.

I had forgotten much of my Spanish when I first returned to Chile and it was fortunate that most of the pilots spoke some English. I was reasonably fluent by the time I was assigned as co-pilot to Captain Lopehandia who spoke no English at all. One of the most senior pilots, he preferred to fly only on the route to Buenos Aires and I made my first flight there with him.

It was one thing to fly north or south with the Andes mountains clearly visible on one side or other of the aircraft. It was quite another to fly directly towards the mountains and cruise through a series of passes at an altitude of 13,000 feet when the peaks extended far higher than that level and Aconcagua, the tallest of them, reached nearly 23,000 feet. Lopehandia soon tired of flying the Lodestar as we made our ascent and indicated that I should take over. Then he took the morning newspaper from his briefcase and became absorbed in its contents. I kept the aircraft on the course he had chosen and levelled out at 13,000 feet. Before long we were flying directly towards what looked to me to be an impenetrable wall of mountains all heavily clad in snow, Aconcagua in all its magnificence clearly visible. Light turbulence rocked the aircraft and to my relief Lopehandia looked over the top of his newspaper and then indicated a valley which offered a minuscule space for an aircraft to pass in an easterly direction.

Possible routes through the Andes had been surveyed and one finally chosen by Mermoz of the French airline Aeropostale in 1929. Mention has already been made of his earlier exploits opening a mail route to French West Africa. In the course of one Andean survey flight in an aircraft with an absolute ceiling of 16,000 feet icing forced Mermoz down on to a flat tableland 12,000 feet above sea level, the sides of which were sheer. The only hope of escape was to run the aircraft off the edge and hope to glide down for a safe landing on lower ground. He did this and survived. Another French pilot, Guillamet, survived a crash just to the south of the Maipu volcano. Having a family and knowing that he would not be legally presumed dead until four years had passed he tried to make himself as visible as possible to searchers. He was seen and rescued before his injuries and frostbite killed him.

Lopehandia put aside his newspaper and directed me along the warren of passes, pointing out special features such as the statue of the Christ of the Andes. This marks the border with Argentina and had been erected to signify the desire of the two nations to live at peace with one another. As we emerged over the Argentine plains he pointed to the town of Mendoza. As I was to discover on subsequent flights pilots bound for Chile who found the passes closed by cloud landed at Mendoza and awaited better weather. At certain times of year the route was closed for several successive days and the disgruntled passengers often resorted to the Trans Andean railway rather than incur hotel charges. The airlines did not then accept responsibility for passengers delayed by weather.

In the fine weather of that season of the year the remaining part of the flight was uneventful. We passed over a few rivers and saw some towns, and the wireless operator transmitted our position reports and obtained a weather report for Buenos Aires. Perhaps uneventful is not quite the right word. During the flight it was obvious to me that the main fuel tanks were nearly empty and I suggested that I switched the selectors to the auxiliary tanks. Lopehandia refused to let me do this. He preferred to let the tanks run dry, the engines to cough and only then to switch tanks. That way he knew exactly how much fuel remained. For nervous passengers suddenly hearing the engines splutter and feeling the momentary loss of power it must have been a heart-stopping experience.

The schedule called for an overnight stop and Lopehandia generously insisted that I should be his guest at a restaurant renowned for their superb beef steaks. Over coffee he enquired how I proposed to spend the evening. I had a number of relatives in the city. My grandfather had been general manager of the British-owned railway and my mother and her sister were both born in Argentina. I replied that I would be calling on friends.

"I cannot persuade you to accompany me, then?" he asked.

"What are you going to do?"

Lopehandia knew at least one word of English and he expressed it in robust fashion although his pronunciation was in error, the 'u' of the second letter sounding like an 'o.'

"There are some very good houses here," he added, relapsing into Spanish.

Whilst I was flying with LAN the British airline which I had originally hoped to join had been developing its route structure, and converted Lancaster bombers began a once-weekly service to Santiago. Carrying thirteen passengers they cruised over the Andes well above the highest peaks but as the cabin was not pressurised oxygen masks had to be worn for that stage of the flight. To avoid discomfort to the ears of passengers the rate of descent had to be restricted to 250 feet a minute. Consequently the Lancastrian spent a considerable time circling down over Santiago before finally landing. A subsidiary of Pan American Airways called Panagra flew down the west coast route in four-engined unpressurised Douglas Skymasters to Santiago and were about to introduce the pressurised Douglas DC-6. The Chilean airline had no plans to introduce more modern pressurised aircraft.

A pilot with whom I had shared lodgings in Southampton had joined the British airline and got in touch with me. We met at the Prince of Wales Country Club where he introduced me to others of his crew. Around us the club members were enjoying the facilities, golf, tennis, cricket or basking in the sunshine by the swimming pool. Attentive waiters were scurrying around with trays of iced drinks.

"This is absolutely idyllic," my friend remarked to me. "What a contrast to England as it is at present. Are there any flying jobs going here?"

"I don't think you would be able to live very well on half your present salary," I told him. "Inflation is pushing up prices of everything all the time. I am thinking of returning home to England to join you."

This statement brought exclamations of astonishment verging on disbelief.

"Do you realise what it is like nowadays back home? The freezing weather has just about shut down industry. Many households are without coal. Food rationing is tougher than during the war. There are no homes to rent and chaps out of the services are desperately trying to find somewhere to live. Someone lucky enough to have a car only gets enough petrol for a few miles. How can you think of leaving a place with a climate like this?"

"Don't be beguiled by the sunshine and the happy faces here," I argued. "I can't afford to be a member of this club. I am drawing heavily on my savings and will soon be spending my war gratuity to pay my food and rent bills."

I was not exaggerating. Chile was a delightful place to live if one was a member of the reasonably affluent middle class. As in England before the war this class employed at least one domestic servant in the house and could always count upon the services of a gardener when required. There was a plentiful supply of caddies for golfers and ballboys for tennis players. LAN acknowledged the fact that they did not pay their pilots sufficient to run a car by arranging for the collection of crew members from their homes and taking them back on their return. Those of us who did not have a telephone could count upon the airline sending a message to our wives if we were delayed by weather on the other side of the Andes.

The actual flying with LAN had been an enjoyable experience. I had been most courteously treated by my Chilean colleagues who never criticised my less than perfect command of their language. The aircraft were very well maintained in most respects and I didn't experience an engine failure while I was in their employ. Nevertheless the salary structure was so dire that I decided that I would have to leave. Two of the former RAF pilots whose families had also been established very many years in Chile decided to stay. They soldiered on until their sixtieth birthdays, when they were flying Boeing 707s. Their final months were spent battling the management for a pension on which they could live.

My opportunity came when I flew to Buenos Aires on the same day that the Chief Executive of the British Airline, Air Vice Marshal Bennett, arrived there. I went to the Plaza Hotel where he was holding a press conference and after its conclusion managed to speak privately to him. He listened to a short recital of my qualifications and asked only two questions.

"Have you had any accidents?"

I had not.

"Will your company object to you joining us? I don't want to be accused of poaching pilots from my competitors."

I did not expect any trouble over that. The route to Buenos Aires was the only one on which LAN met any competition. The company had a monopoly on the domestic routes. Moreover they had not had to spend anything on training me. Bennett hired me and I was authorised to fly home as supernumerary crew when I had worked out my notice. Bennett made it plain that I would have to pay the fare for my wife and stepdaughter. I decided to keep the lease on my flat and sublet it furnished as I would not otherwise recoup all the money I had spent on furniture. The British Embassy rented it for their archivist.

There were some tiresome formalities before I could leave Chile, obtaining documents from the income tax authorities confirming that I had paid my dues, from the police that I was not wanted for any crime, from the armed forces that I had no obligation to perform military service. Visas had to be obtained from the Argentine consulate despite the fact that the Lancastrian only made a refuelling stop in that country on the way to Britain.

I developed a bad cold in my last week before leaving LAN but did not like to plead sickness at that late stage so set off with Lopehandia for Buenos Aires. The schedule had been changed. After refuelling there we were to pick up passengers and fly back to Santiago. Inevitably the heat of the sun on the parched earth would cause strong updraughts, particularly over the Andes where we could expect the Lodestar to be rocked by turbulence. A cold front had been forecast to affect the route in the latter part of that day and I thought it likely that the passes would be closed and that we would have to stay overnight in Mendoza.

I was right up to a point. I had not foreseen Lopehandia's determination to make the flight back to Santiago on that day. Although the Andes weather station reported the passes enveloped in cloud Lopehandia flew along the first of them until increasing darkness and blinding rain convinced him that there was no way through. Then he headed south climbing to over 20,000 feet while the crew put on oxygen masks and the passengers sucked on their tubes of oxygen. When he altered course to the west we were heading for a gap between two enormous mushrooms of cumulonimbus cloud from which, even in the bright sunlight at that altitude, vivid stabs of lightning could be seen. The air became extremely turbulent but it was possible to make the crossing without flying into cloud and as conditions became calmer Lopehandia headed north towards Santiago. However not for him the gentle descent which would have spared the discomfort to everyone's ears and prevent agony to my own with my blocked nose. As I watched the rate of descent indicator touch 700 feet a minute I pointed at it, then to my earphones.

"Takes too long, very boring," he said and laughed uproariously, an act which probably helped to clear his own ears.

That was my last flight with LAN and a few days later we were on our way home in a Lancastrian. On the first sector to Buenos Aires all thirteen passenger seats were occupied but there was an oxygen mask in the solitary lavatory in the tail so I sat there for the Andes crossing. Luckily the air was calm. Passengers caught there in severe turbulence suffered the most unsalubrious experiences. The flight to Britain was on schedule and uneventful. As we circled over London airport snow was visible over the ground. The big freeze of the winter of 1946 had extended its grip into March of 1947.

The author being awarded his wings at RAF Penhold, Alberta, during the winter of 1941 (via A.S. Jackson)

The flight deck of a "C" Class Short Empire flying boat (Don Munro via A.S. Jackson)

BOAC 'Hythe' Class Sunderland III over the pyramids (via A.S. Jackson)

Short S.26 "G" Class flying boat *"Golden Hind"* after its return from the RAF to BOAC, showing the 'Speedbird' symbol. It flew its last service in 1947 (via A.S. Jackson)

The Chileans ordered Junkers 52s & 86s in 1938 but they were withdrawn during WW2 owing to spares shortages (via A.S. Jackson)

LAN Chile bought Lockheed XIV Lodestars during WW2. Alongside is the DH Gipsy Moth which had inaugurated the domestic air mail service in 1929 (via A.S. Jackson)

Rangoon 1945: flight crews of 267 Dakota Squadron. The author, without cap, is standing fourth from right (via A.S. Jackson)

Heathrow as it was in January 1946 (via A.S. Jackson)

BSAA Lancastrian at Limatambo Airport, Peru in 1947
(family of AVM Bennett via A.S. Jackson)

January 2nd 1946: BSAA Lancastrian *"Starlight"* about to make the first flight from Heathrow. AVM Bennett (with briefcase) stands next to Capt. Cracknell DSO, DFC (via A.S. Jackson)

A DC-3 of LAN Chile; note the unusually enlarged fin (via A.S. Jackson)

Glenn Martin 202; not used for long by LAN as prop tips had inadequate ground clearance on uneven landing surfaces, and US crashes with the type hastened replacement (via A.S. Jackson)

An Avro Lincoln in the colours of the Argentine Air Force, flown by the author to Buenos Aires in 1948 (via A.S. Jackson)

Passengers boarding a BSAA Tudor IV in 1948 (via A.S. Jackson)

BSAA Avro York *"Star Leader"* in 1946 (via A.S. Jackson)

BSAA Lancastrian *"Star Trail"* is prepared for service to Chile. In September 1947 it was damaged beyond repair landing in a severe storm on Bermuda (via A.S. Jackson)

Passengers boarding a BOAC Argonaut at Cairo Airport (known as Farouk) in 1950 (via A.S. Jackson)

The north side of London Airport (Heathrow) in 1953, prior to construction of the central area (via A.S. Jackson)

An evocative shot of a BOAC Britannia at Entebbe's old terminal in 1958 (via A.S. Jackson)

The last type flown by the author: a BOAC VC-10 (via A.S. Jackson)

CHAPTER 8

THE STARLINERS

At the time I had written from Burma to the airline which was proposing to inaugurate services to South America it had been a private concern and had not asked for subsidy from the government. After the Labour Party won the general election of 1945 the airline was nationalised and became British South American Airways Corporation. The original board of directors were permitted to remain in office.

The only one of these with any experience of aviation was Air Vice Marshal Bennett who had distinguished himself during the war when he had commanded the Pathfinder Group of Bomber Command. Prior to that he had been a Captain in Imperial Airways and the founder of the North Atlantic Ferry Organisation responsible for the delivery of aircraft from American factories to the United Kingdom. The other directors were the representatives of the five shipping companies which had served South America over many decades. The Chairman of the Board was John Booth whose vision it had been to complement their maritime routes with links by air.

The airline's head office was in a modest building in Grafton Street off Piccadilly and the base for the aircraft was Heathrow, with engineering and maintenance facilities also available on a small grass airfield at nearby Langley. Bennett had engaged men who had served under him during the war to form the nucleus of his management team. These included two former Imperial Airways pilots, Gordon Store and David Brice, as operations manager and chief pilot together with the Pathfinder Group's chief of engineering. He had selected a dozen of his former squadron commanders to be Captains of a fleet of Lancastrians and Yorks.

The Lancastrians had the advantage of a ceiling of 25,000 feet or more but normally cruised at 10,000 feet which obviated the need for a continuous provision of oxygen. The Yorks were powered by the same Rolls Royce Merlin engines and the wider body allowed twenty one passengers to be carried but they were slower than the Lancastrians and had a shorter range. Their performance was also inferior. If for any reason an outboard engine was shut down the pilot had to be quite certain that he was going to be able to land after selecting full flaps because the York could not be kept on a straight course if full power was applied to the three good engines. Passenger services to South America had begun in 1946. When I joined there were three York services each week to the east coast of the continent. One Lancastrian each week flew down the west coast to Chile landing en route in Portugal, the Azores, Bermuda, Jamaica, Colombia and Peru. All the aircraft were given names prefixed by the word 'Star' and our call sign was 'Starline.'

There was a legal requirement that commercial airliners flying over the oceans should include among the crew a navigator holding the first class licence. Before the war very few airmen had gained this qualification and to allow British airlines to develop their network of routes the regulations had been relaxed as a temporary measure. At the conclusion of hostilities BOAC had re-established quite a number of routes and this had been made possible by the large numbers of pilots, navigators and wireless operators seconded from the RAF. Subject to their obtaining the civil licences many of these were offered contracts.

Bennett had a deep-seated prejudice against BOAC dating back to the time Imperial Airways had been nationalised and merged with another airline to emerge as a state corporation. He had no intention of employing navigators, apart from two instructors, preferring to enrol pilots who were required to obtain the first class licence. Having achieved this, promotion to Captain followed fairly swiftly in the first year or so of the Company's existence. Having spent my war service mainly in the East where decorations were rarely awarded I was most impressed by the array of medal ribbons worn by the Captains. Whatever our former rank those of us engaged as Second Officers wore one single ring round the sleeves of our jackets.

Wireless operators were also required to obtain civil licences and Bennett was fortunate to obtain the services of some well-qualified men who had spent years at sea before transferring to aviation. It was their task to bring up to standard ex-RAF operators who had often been expected to combine that function with duty as an air gunner.

No airline crew would be complete without a stewardess to attend to the passengers and in BSAA these were known as 'stargirls.' The first one to be employed, Mary Guthrie, had been an ATA pilot. Almost without exception the other girls had been in the services and a few of them spoke Spanish or Portuguese, having volunteered during the war. It was not the policy to recruit married women or divorcees, but there were several widows of RAF pilots. All were expected to resign before reaching their thirtieth birthday. No stewards were employed.

It was necessary for me to have my 'B' licence endorsed as competent to fly the Lancastrian and York. The formalities were fairly perfunctory. Six landings performed satisfactorily in daylight were deemed to be adequate in addition to passing a written examination comprising about fifty questions. This was basically an exercise of one's memory. The fuel system showing the tank capacities had to be sketched together with the oil system. The formula for loading the aircraft within its centre of gravity had to be remembered together with the location of safety equipment such as fire extinguishers, axes and life rafts. The bulk of the questions concerned such matters as the maximum and minimum oil pressure and temperature, tyre pressures, the maximum airspeed below which sections of flap could be extended and so forth. One knew what most of the questions would be and it was perfectly possible to pass the written examination without having seen or set foot on the aircraft. Most pilots passed it first time. Regrettably I did not, having reproduced from memory an inadequate drawing of the fuel system.

I had only been back in England a few weeks when I had the good luck to be offered an unfurnished flat in an area conveniently close to Heathrow. I had been making enquiries at the offices of estate agents without any real hope of success when I met one whose brother-in-law was BSAA's manager in Buenos Aires. Someone had just vacated a flat and when I mentioned my connection with the airline I was offered the tenancy. It was on the top floor of a building and coal had to be carried up three flights of stairs. The roof leaked but fortunately not into all the rooms. It was possible to travel by bus to the airport in about twenty minutes. I counted myself very fortunate indeed.

My licence legitimised I was soon on my way back to South America. Among the crew I was not the only new member. The Yorks carried two stewardesses and a new one did not wear a uniform, the big freeze of the winter having put the outfitters temporarily out of business. As each passenger was seated he was offered cotton wool to protect his ears from the fearsome roar of the engines and boiled sweets to suck during the climb and descent.

As the chocks were removed from the aircraft wheels I observed the routine ceremony which was practised. The civil air ensign and the Union Jack had been hoisted by the crew on short masts which protruded from the cockpit. The staff who had attended the departure, the engineers and receptionists stood in line alongside the traffic officer who saluted smartly as the aircraft was taxied away.

It was customary for the Captain and the other two pilots to take it in turns to act as navigator. The oceanic sector, normally flown at night when position fixes could be obtained from star sights, was navigated by the Captain unless one of the other pilots had proved his competence to keep an accurate plot. Bearing in mind that even an accomplished navigator could do no better than plot on his chart the aircraft's position twenty minutes earlier the importance of this job was obvious. Fortunately the winds at 10,000 feet in the South Atlantic are never very strong so no one drifted very far off course.

I was allowed to navigate from Lisbon to Dakar.

"I don't think you can get us lost!" the Captain remarked. "We cross and recross the coast of Africa and the rotating beam of the lighthouse at Dakar can be seen when we are ninety miles or so away. Just as well," he added. "There is nothing else in the way of aids there: no radio beacon, no approach lights to the runway, nothing."

On the flying boats in 1942 we had remained to the west of the land mass of Africa until we had arrived abeam Gambia. That was no longer necessary and there were a few ports which served as landmarks to assist navigation. The route also offered a good opportunity to practise with the sextant as it became dark soon after leaving Lisbon and the flight took eight and a half hours.

On a visit to the galley during that flight I found the new stargirl being warned by her colleague about the tricks the crew were liable to play on her.

"They will ring the call bell and order you to bring them the golden rivet," she was told: "there is no such thing. They will invite you to look out of the window at something interesting and slide the window open for you to have a better view. Then there will be a terrific draught and your skirt will blow up above your waist. The navigator will ask you if you would like to see the new moon through the sextant. When you return through the cabin all the passengers will laugh because the eyepiece has been smeared with shoe polish and you will have a black eye."

When she had finished describing what she called "these childish antics by moronic oafs" she confessed to me that she had on one occasion offered appropriate warnings to a male member of the crew whom she had assumed to be a recent recruit owing to his lack of a uniform.

"He seemed rather surprised and began asking me a lot of questions about my job and the problem I had to contend with. I was most embarrassed when I discovered that I had been offering advice to the Air Vice Marshal."

Bennett often flew as a member of the crew and as he was a youthful-looking thirty-four-year-old when BSAA was formed her error was easily understandable.

Dakar was not a popular station for night-stopping crews. The dingy hotel would not have rated any stars in Britain. Three of us shared a bedroom and we were luckier than other guests who slept in the passages as accommodation was so scarce in the town. The choice of food was very limited and omelettes featured on the menu at most meals. However the company had agreed that we could order bottled mineral water as the tap water was suspect. Bottled beer was cheaper than this and the hotel manager was happy to serve all we required so long as we signed for it as mineral water. It was some time before a visiting accountant arrived to investigate the high consumption of this item. Separated by over two thousand miles from the government of Vichy throughout the war and despised by their former allies the resident French were not particularly friendly to us.

Another York would arrive from England soon after midnight. For the passengers there would be two further sectors to be flown before they could hope to sleep in a bed again. The sector to Natal in Brazil took about nine hours and after refuelling a further seven hours passed before landing at Rio de Janeiro.

For many passengers this would be their first flight in an airliner, air travel not having been a widely shared experience before the war. Some were only on board because they had been unable to obtain a passage on a ship. Many years were to pass before businesses began sending any but their most senior executives by air. We received complaints about the ban on cigars and pipes and the temperature of the cabin.

"Let's have a window or two open," passengers were heard to say.

Those accustomed to travel by ocean liner remembered that it was customary to tip the cabin stewardesses and others at the conclusion of the voyage. Among other airlines BSAA strictly forbade this practice and a notice in the aircraft lavatory declared that "the staff of the company do not expect nor are they permitted to receive gratuities of any kind." The stargirls received very low salaries, partly because theirs was thought to be a glamorous job and the company was never short of applications to perform it. It was no easy task to live on the salary and one girl had already been sacked for removing the notice. It had been her practice to replace it when the passengers had disembarked at Heathrow. The day came when she forgot to do so and the notice was found hidden in the galley.

The Radio Officers were paid less than the Second Officers and Captains were no better off than Flight Lieutenants in the RAF. Our Radio Officer, a single man, was living in lodgings near the airport.

At first he had not been invited to share the living room with his landlady's family. He told me he had adopted an ingenious stratagem to make his conditions more comfortable. When the family were out he went into the sitting room and put their wireless set out of action. Later there was a knock on his door and as he had expected he found his landlady begging his assistance.

"Yes, of course," he said and was soon feigning a close examination of the set. "Oh dear me," he remarked; "you are in real trouble here, no doubt about that; might cost you a bomb in a repair shop. I had better take it up to my room and see what I can do."

The next day he returned it in working order but with a warning that the fault might well recur. He was invited to make use of the sitting room whenever he liked. It occurred to me that other electrical equipment could also need attention but perhaps he was now perfectly comfortable.

He shook his head. "One can have too much of a good thing. They have bought a TV set and drop heavy hints about how lucky they are to have me as a lodger. Their spotty daughter wants me to bring nylons and other scarce things back from abroad for her." He sighed. "I think I may have to find another place."

In the morning we landed at Natal, an aerodrome which during the war had been used as a ferry point for hundreds of military aircraft on their way to Africa. Now it had a deserted look although in addition to BSAA the aircraft of Air France, Iberia and KLM also called in for fuel. The South Atlantic had first been flown as a commercial venture by the French and the redoubtable Mermoz made the crossing in 1930 in twenty one hours in a seaplane. On the first occasion on which he experienced an engine failure he put it down in the ocean alongside a steamer. In 1936 he reported engine trouble again but no trace of his aircraft or his crew could be found.

Brazilian stations from which we obtained weather reports and which received our routine position reports used a 'bug key' to transmit morse at a very fast rate. It was beyond the ability of a few of our Radio Officers to copy their messages. It did not help them that the aircraft's wireless equipment was obsolescent if not actually obsolete. By no stretch of the imagination could it be described as modern. One Captain was extremely dissatisfied with his colleague's performance and the latter, during the night stop, poured out his troubles to the King's Messenger, a former Wing Commander under whom he had served during the war. The latter found the opportunity to have a chat with the Captain and asked for his forbearance.

"That chap was the wireless operator of the only crew to return from one of my squadron's raids," he revealed. "The squadron took fearful punishment. He had done his fair share and I posted him away for a rest. I don't think he can have had much recent operating practice when he joined your lot."

"What did he do after you posted him then?"

"Oddly enough he was given an admin job in a military prison."

Conversely useful assistance was sometimes available from the most unlikely persons. We used a radar system called Eureka Rebecca which enabled the navigator to measure the aircraft's distance from a similarly equipped airport when it was flying within a ninety mile range. The stargirl collecting dirty cups and saucers from the flight deck heard the navigator complain to the Captain that his set seemed to be unserviceable.

"Can't get a peep out of the bloody thing," he grumbled, thumping the top of his receiver in disgust.

"Let me have a look," the stargirl said. "I was a radar mechanic in the WAAF." Ten minutes later the set was in working order.

Rio de Janeiro's airport was closed by fog when we arrived in the area so a diversion was made to the military airfield of Santa Cruz further to the south. A hangar built for the Graf Zeppelin was visible there. The Captain and I were taken to meet the Base Commander.

Offering us coffee he enquired: "Would you like to see some pictures?"

Rather tentatively the Captain accepted and our host enthusiastically produced a thick volume which, he told us, was a record of the Brazilian Air Force's contribution to the allied war effort in Europe. This was news to us, unaware that any Brazilian units had seen any action, but we were shown pictures of pilots and aircraft in Italy. Many of the photographs were of airmen who had died as a result of accidents. It was a relief to us when our examination of the album was interrupted by the information that the fog had cleared at Rio and we could resume the flight.

The magnificent bay surrounded by hills and dominated by the huge concrete statue of Christ was free of any cloud cover when we arrived overhead. The Captain pointed out to me the famous Sugar Loaf mountain at the entrance to the harbour and the two airports, the smaller of which was used by the domestic carriers.

"When the cloud base has been down to a few hundred feet some Captains have flown in low through the harbour mouth," he told me. "The danger you face if you do that is from the risk of collision with the masts of all the ships."

Customs clearance was swift and courteous and it was explained to me that all the international airlines paid the Customs officers 'sweeteners' to avoid harrassment of the passengers. These officials were very poorly paid. At airports such as Natal the airlines had to send cars to bring them to work or the delays to flights would have been horrendous. In my innocence I was rather shocked by this revelation.

"Do you mean to tell me that the Customs Officers call at your office for their bribes at the end of each month?" I asked the local manager.

"No! They want their money in advance at the *start* of the month."

If that surprised me what distressed the passengers was the state of the lavatories in the airport buildings which were under military control. Some of them believed that it was the airline's job to do something about it. They complained bitterly to the Company's representatives most accessible to them, the crew. Just as the prevalence of litter and uncollected refuse dismays visitors to Britain so the lavatorial habits of South Americans repel northern Europeans.

The night stop in Rio de Janeiro was very welcome to both passengers and crew. The following morning the flight continued to Montevideo and Buenos Aires where another comfortable night was spent. The crew of the York then boarded a Lancastrian to conclude the outward journey at Santiago.

There was no public address system in those days to keep passengers advised of the progress of the flight or to draw their attention to points of interest visible from the windows. Bulletins were sent back to them, "crossing the line" certificates issued and frequent visits to the cabin were made by the Captain and co-pilot to build up goodwill and engender confidence.

Not all of the passengers liked to see the Captain away from his post and made it clear that they would prefer him to stay on the job. A few even expressed surprise that the same crew did not operate the aircraft without rest all the way from London to Buenos Aires.

"Do you mean to say you are paid for sitting around the hotel swimming pool?" was the sort of remark one became accustomed to hear.

Other passengers tended to regard pilots as the source of all wisdom.

"I need your help," a man confided to me in a low murmur as passengers and crew checked in at the hotel in Buenos Aires. "My wife is getting rather worried."

He jerked a thumb towards the lobby where I noticed at once the staring eyes of a young woman looking nervously in our direction.

"It's been five days now and nothing, you know, nothing."

"I am afraid I don't follow," I said, fearful that I was about to be made privy to some intimate family problem.

"Constipation is my wife's problem," he explained. "Where can we get something for it?"

Grateful for small mercies I accompanied him to a pharmacy a short distance from the hotel.

On their first visit to Buenos Aires crew members were astonished at the range and quality of goods on offer in the smart stores along the Calle Florida which bore such names as Harrods or Gath and Chaves. In Britain rationing of food and clothes, shortages of household goods, bicycles and motor cars lasted for years after the end of the war. To prevent a breakdown in the national economy government regulations put severe restrictions on overseas travel by controlling the amount of currency British residents could take abroad. In the case of crew members the limit was two pounds. It was illegal to cash a cheque in a country outside the sterling area which included most of the world except the British Commonwealth.

We wanted to buy presents for our long-suffering wives and canned foods which were difficult to find in Britain or other things which would brighten up our homes. One method was to accept local currency for a cheque which was posted to England for encashment by a resident or taken there by a returning traveller. More ambitious individuals did a little trading on their own account. They had observed that in Dakar high quality French perfumes were inexpensive whilst the same goods fetched a high price in Brazil. The perfume was taken in a bag to Rio where the aircraft remained overnight. A member of the ground crew took charge of the bag whilst the crew cleared Customs and took it to the hotel. This stratagem provided some spending money after the sale of the perfume to a shop in the town. It came to an end when a baggage handler thought the bag had been overlooked and obligingly carried it into the Customs shed. The crew, including one man whose face visibly turned pale, denied all knowledge of it. A Customs Officer opened the bag and someone whom it would be kindest to identify as George left shortly afterwards for pastures new. We were to see more of him later.

More openly and to the indignation of the Company a new stargirl on her first trip to Buenos Aires bought her wedding trousseau and other finery and on her return here promptly resigned. She had been a member of my crew and subsequently I overheard a heated argument between two stargirls about her conduct. One of them defended her.

"She had a lousy first trip. On the way south there was a lot of bad weather and half the passengers were airsick. One old lady lost her false teeth and the senior stargirl made her search and recover them. If that wasn't bad enough she had to contend with some drunken Uruguayans on the way north who pinched her bottom black and blue. Not very pleasant things to happen to a decent girl on her first trip."

The other girl would have none of it.

"Excuses, excuses!" she retorted. "You were taken in by her innocent vicar's-daughter look. She had it all planned from the word go. Her wedding invitations were already printed."

Most of the girls did continue flying for a number of years, ready to take the rough with the smooth and undeterred by the less than chivalrous attitude shown to them by some of the male crew members.

"Oh! my God, have we got you again?" I heard one Captain say to the stargirl as the crew assembled at Heathrow. "Why can't the airline take on some fresh young girls?"

"What do you suppose it's like for us?" came her reply. "You lot are not exactly oil paintings."

Sitting on the bench at Dakar one day I did notice a stargirl whom I had not seen before.

"This must be your first trip," I said to her.

"Yes it is. How did you guess?"

"By your prewar bathing suit. The stargirls all buy themselves new ones on their trip to Buenos Aires."

I was instantly ashamed of this gaffe as she blushed and put a hand over a crop of moth holes. We became good friends and she liked to remind me of my crass introductory remark. Her name was Lennie Will. Crew members had their own luggage tags on which we wrote "northbound" or "southbound" to try and prevent misdirection by loaders. Some joker reading the words "Lennie Will northbound" added a few more: "Lennie wouldn't southbound."

CHAPTER 9

DISASTER IN THE ANDES

Before long I made my first journey from Heathrow on the mid-Atlantic route to Bermuda. The first stage was to Santa Maria in the Azores and took about seven hours. After refuelling we continued to Bermuda and this could take anything up to fifteen hours. At that time this was the longest oceanic sector flown by any airline in the world. There were no weather ships at that latitude to monitor and report the upper wind strength and direction. Very few other aircraft operated in that area. As a result the meteorology station at Santa Maria was unable to provide accurate information on which a flight plan could be prepared. An added complication was the absence of any airport which our Lancastrian could reach if Bermuda was found to be closed by bad weather. This was to prove a major problem when the Avro Tudor with its shorter range than the Lancastrian was introduced on to the route.

The crew always felt a surge of relief when the lighthouse on Bermuda came into view, concluding a period on duty of about twenty two hours. Kindley airport had been built by the Americans in 1940 when the United States were offered a base on the Crown Colony in exchange for fifty destroyers which had been mothballed since the first world war. Life was never to be the same again for the islands which could previously only be visited by sea. Not even the Governor of the colony had been allowed a motor car. In 1947 the USAF continued to keep aircraft there including a squadron to track the movement of hurricanes, together with a Search and Rescue unit.

We were accommodated at The White Horse Tavern in the Parish of St George. This pleasant old-fashioned inn was owned by a retired English seaman called Arthur and his wife Rosie. Crew members shared bedrooms and there was a bathroom on the landing. In the morning we entered the kitchen to tell the cook what we wanted for breakfast and helped ourselves to fruit juices and cereals. Arthur would always cash our cheques as Bermuda was in the Sterling Area and he bent the rules by providing us with dollars which was the preferred currency for shopping in South America. It happened that my youngest brother was posted to Bermuda with a company of the Gloucestershire Regiment in which he was a national serviceman. Arthur remembered him at Christmas, generously inviting him to spend it as his guest.

The White Horse Tavern was no Hilton but we liked the atmosphere and it came as a nasty shock when we heard that our management thought the charges were too high. Near by there was the shell of an old property which had housed German prisoners of war. Someone had proposed that it could be turned into crew accommodation. We went and looked at this site in gloomy apprehension but the project was not pursued.

Economy was at the forefront of the mind of Air Vice Marshal Bennett. He had been a Captain in Imperial Airways when in 1939 it had been merged with another company to become BOAC. He had strongly disapproved and continued to regard the senior corporation as bloated with unproductive staff and thoroughly inefficient. He wanted no idle hands in BSAA and between trips overseas there were jobs for all of us to do. This included air tests before scheduled flight departures, the checking of compasses after engines or other components had been replaced and the preparation of navigation kits for aircraft. The stargirls were found work in the catering department including scrubbing the floor.

The lengthy periods of duty when flying owed something to the fact that no legislation existed at that time to control the situation. Virtually everyone engaged in the burgeoning number of airlines, large and small, were fit young men whose service training had not inclined them to question the length of time which they spent at work. Bennett himself had shown that he was capable of astonishing periods at the controls when in 1938 he had flown the seaplane *"Mercury"* to Quebec and immediately thereafter to New York with only a wireless operator as companion. He had followed that with his record-breaking nonstop flight from Dundee to the Union of South Africa, a journey of over 6,000 miles.

Bennett had no objection to his pilots taking it in turns to curl up under the navigation table and try to sleep. He kept a tight rein on the level of staffing and was delighted to find that the wives of staff overseas enjoyed accompanying their husbands to greet passengers in transit at some unearthly hour in the morning, particularly as they were not on the payroll.

The company had suffered the first serious accident prior to my return to England when an aircraft being flown by a Captain on his first command crashed on take off at Bathurst in West Africa. Everyone on board was killed. It transpired that the pilot had less than a dozen hours in command of a York and had never previously taken off this type of aircraft at its maximum weight. In 1946 no aircraft operating manuals recognised the fact that airports which experienced very high temperatures (and airports some thousands of feet above sea level) markedly degrade take-off performance.

Before long another Captain on his first command was descending towards Dakar just as fog was beginning to envelop the airport. It had not been forecast and there were no facilities for an instrument approach, no illumination other than the feeble runway lights. At the airport the Captain who was waiting to take over realised the danger and tried to persuade the controller in the tower to set fire to barrels of tar or other inflammable material near the runway threshold. This was beyond the controller's competence. He was in a panic and tried to persuade the York Captain to divert to a place called Thies. This was despite the fact that he could not communicate with Thies nor was its runway lit. Thies did not figure in our route book and the Captain made an unsuccessful attempt to land at Dakar. There were injuries and some fatalities when the aircraft overshot the runway.

During 1947 I obtained the First Class Navigator's Licence and looked forward to a command. The policy was to start us off with some trips on which only cargo was carried and I duly flew out of Heathrow for Buenos Aires on a freighter. A problem arose when I asked my Second Officer to pass me the route book so that I could refresh my memory on the radio frequency beacons in the Lisbon area where the first landing would be made.

"Route book?" he answered. "No route book among our stuff. They are a personal issue to Captains."

This was true and in the circumstances of this trip I should have remembered that and made sure that the duty crew at Heathrow had produced one for me. Fortunately the weather at Lisbon was good. At Dakar I borrowed a Captain's route book and laboriously copied out all the information I thought I might need for the rest of the trip.

The next time I flew as an acting Captain was to deliver a Lincoln bomber to the Argentine Air Force. They had purchased eight of these from the RAF. I had to collect the aircraft from an airfield near Grantham and upon examination it appeared to me to be in every respect a replica of a Lancaster. The test pilot who gave me a circuit did not disillusion me, nor did he have available any book of instructions for operating it.

The more junior of my two Second Officers was as well qualified with licences as I was but had accepted the job at only half the normal salary. He had been told that BSAA had a full complement of pilots but he could start as an apprentice until there was a vacancy. Grant-Jones was a Welshman, married and with two children, whom we called "Taffy."

There was only one wireless telegraphy receiver in the Lincoln and this meant that it could not be used for obtaining bearings by the navigator when the Radio Officer was using it to obtain route clearance, pass position reports or to request weather information. But when we were flying along the coast of Brazil and were in voice contact on the VHF equipment Taffy had free use of the W/T receiver and kept me advised of our progress.

"We are now abeam Caravelas," he would announce and later: "now abeam Vitoria."

To which my co-pilot Basil Milsom always replied: "Taffy, you are worth every penny of your four hundred pounds a year."

Another pilot waiting for his command was Lincoln Lee. He was offered the opportunity to fly a specially-equipped Lancaster from Heathrow non-stop to Bermuda. On the way he would be intercepted by an aircraft tanker based in the Azores which would replenish his tanks. The principle of flight-refuelling commercial aircraft had first been tried in 1937 when Bennett had participated in a series of flights across the North Atlantic on flying boats. The war interrupted the trials. Lincoln Lee swallowed his astonishment.

"Certainly," he said, gulping hard. "As it happens I have not yet flown to Bermuda and don't know the first thing about the technique of flight refuelling. I will take it on if time can be found for me to do one or two night landings on a Lancastrian. Up to now just about all my flying in the company has been on the York."

"You don't sound too keen," he was reproved. "There are other chaps hoping to be given a command soon."

Lee made the flight and completed the return trip successfully also. I delivered my aircraft in one piece to the Argentine and when I arrived home went along to the chief pilot's office in Grafton Street for the post-flight conference, an unpopular chore abandoned decades ago.

"No problems I imagine?" he remarked. "Did you bring back the maintenance log?"

This was the book in which a crew member recorded each hour all the instrument readings. I had regarded it as the property of the purchasers of the Lincoln and had handed it over. I had explained that and mentioned that I had expected the aircraft's performance to be superior to the Lancaster but I had noticed no difference.

The chief pilot looked surprised. "Really? What power settings did you use for the cruise?"

When I replied that I had flown it in every respect as if it was a Lancaster he was even more surprised and indignant that the people at Grantham had not supplied an operating manual or given me a thorough briefing.

I was not too surprised myself because it was so typical of the manner in which aviation was being conducted at that time. John

Gilbert had returned to the RAF soon after being demobilised because he could not afford to support his wife and two children on a law student's allowance. He had flown two tours of operations on Halifax bombers and Flying Fortresses. Reporting for duty he was astounded to be ordered to ferry a Spitfire to another station.

"The only single-engined aeroplane I had ever flown was a Tiger Moth," he told me. "I took off in the Spitfire and made the short journey thinking that if I bent the thing on landing that would put an end to my recently resumed career."

The news that the Lancastrian *"Star Dust"* had failed to complete a crossing of the Andes en route to Santiago was another blow to our airline. This too was being flown by a Captain on his first command. We had all been warned not to enter cloud over the mountains as the turbulence and icing posed too great a threat. *"Star Dust"* was never found and the wreckage is probably still embedded in one of the permanent glaciers. Aircraft had come to grief in the Andes before and particularly during the winter months no one ever expected to recover survivors alive. Local attempts mounted on both sides of the mountains to look for the wreckage were abandoned after a few days.

Bennett was not prepared to accept this gloomy prediction and flew out to Buenos Aires to conduct his own search. He gathered together a crew of which I was a member and some local employees to keep watch from the cabin windows and we set off in a Lancaster freighter. Five fruitless days were spent on the search and there were periods of considerable alarm as Bennett flew up valleys covered in cloud until we seemed to be headed for an inevitable collision with a mountain. Each such sortie concluded with a very steep turn to fly off and try some other avenue of possibility. Finally he decided to take off from Santiago at dawn, make one last attempt to find the lost aircraft during the hours of daylight and then return to Buenos Aires.

Bennett was well-known for his dawn starts and we were aware that our hotel would not provide us with even a continental breakfast at the very early hour before we were to be picked up. He had not told the local staff to put any provisions on board for us. We saw an opportunity to remedy this when we arrived at the airport and found the station engineer trying to cure an oil leak. Nervously someone suggested that we might be allowed to have coffee in the terminal building.

Bennett exploded. "All you people think about is your stomachs! Have you no concern for those poor people in the mountains?"

To my surprise our stargirl responded in an equally assertive manner: "They have all been dead for days. *We* are the ones who are threatened with starvation."

This outburst seemed to have the desired effect. Bennett conceded that we might have a coffee.

"Be back here in fifteen minutes." He himself remained with the station engineers.

We hastened to the cafeteria and ordered a rapid service of "café completo" and waited with increasing anxiety for the expected coffee and bread rolls.

Bennett suddenly put in an appearance, sat down with us at our table and remarked: "not finished yet?"

At that moment the waiters arrived bearing steaming jugs of coffee, bread rolls and – although not ordered – plates of eggs and bacon.

Bennett leapt to his feet. "I am having none of this. We are off!" Pursued by angry shouts from the waiters we departed reluctantly for a final day among the mountains.

Bennett was often heard to say that he never asked anybody to do anything which he was not prepared to do himself. That was quite literally the truth but overlooked the fact that he was an exceptionally gifted man in a number of fields. In addition to his talents as a pilot and navigator he was able to understand and digest at first reading the most abstruse technical publications. He believed that an aircraft Captain should be free to decide whether to take off and whether to land, unshackled by regulations and without interference from the managers sitting in offices.

After he had appointed David Brice as chief pilot in 1945 the latter was preparing to set off on a proving flight to South America, taking with him some of the airline's future Captains. He went to see Bennett to discuss the route to be adopted and the places where landings should be made.

"You are the Captain," he was told. "You decide."

"Yes, but surely someone has done some costings on the possible routes, compared fuel uplift prices at various stations, varying landing charges which will affect the overall expenditure. Not all the airports have agreed to allow us to pick up passengers."

Brice had held a command for about six years at the time and amassed about 5,000 hours. Some of BSAA's new Captains had under 2,000 hours and heroism under fire over Germany was no substitute for experience of the very variable weather conditions across both sides of the Atlantic ocean, the inferior quality of the runways, approach aids, weather services and air traffic control. Unhappy about the management of BSAA he resigned after a year.

CHAPTER 10

"CAN ANYONE SEE BERMUDA?"

As our competitors began to introduce pressurised Douglas DC-6s and Lockheed Constellations on our routes it became increasingly evident that we could not compete using the aircraft we were flying. Great hopes were placed on A.V. Roe's Tudor which had originally been ordered by the government in 1943 as a suitable contender for the North Atlantic route. It was typical of the way civil aviation was treated at the time that BOAC were given no say over the matter and the manufacturers were prohibited from having any dealings or consultations with the intended customer. When BOAC learnt about the Tudor and found that it had been designed to carry only twelve passengers they made it clear that it would be hopelessly uneconomic and uncompetitive and that they didn't want it. However a prototype had flown in 1945 and others were on the production line. A.V. Roe offered to extend the fuselage by six feet and to accommodate twenty eight passengers. BOAC were put under tremendous pressure by the government to evaluate the Tudor but after tests in the heat of Khartoum firmly rejected it.

Bennett let it be known that BSAA might help to solve the government's problem if the price was right. In August 1947 A.V. Roe's chief designer Roy Chadwick was killed at the controls of a Tudor which crashed on take-off. The cause was due to the reversal of the aileron controls by a careless mechanic. At the time I was with Bennett's crew on the search for *"Star Dust"* and I read the news to him as printed in the local newspaper. He was shocked and disappointed but insisted that the manufacturers of the very successful Lancaster bomber must be capable of building a good airliner. The following month he set off for Jamaica in the lengthened version known as the Tudor IV and confirmed that he would accept the aircraft. That same month he commanded a proving flight to Santiago on which fare-paying passengers were carried. Shortly afterwards the Tudor went into service on the mid-Atlantic route to Cuba, landing in the Azores, Bermuda and Nassau.

I arranged to take the technical examination on the Tudor before I had set foot in one and passed it without difficulty. As a result I was rostered to navigate on a Tudor service to the Caribbean. Sitting at the navigator's station and preparing my charts I realised that I was also responsible for handling the pressurisation panel and controlling the cabin heating system. This was a consequence of Bennett's decision not to carry a flight engineer. An obliging member of the company's engineering section gave me a few brief verbal instructions.

"Remember," he concluded his remarks, "when the aircraft doors are closed you must press your foot on that pedal on the floor there. That will seal the pressurisation."

The traffic officer had brought the load sheet for signature and departed from us. Then the stargirl entered the flight deck and spoke to the Captain. As she turned to leave I asked if everybody was now on board.

"Yes," she said and paused to put a handful of barley-sugar on the navigation table from a tray containing packets of cotton wool and boiled sweets. "I am about to fling these to the passengers."

I pressed the pedal and turned my attention to my more familiar duties. The captain started one of the engines and moments later a flustered traffic officer reappeared.

"Please," he bleated, "I can't get out. The main passenger door simply won't open."

The captain gesticulated to the ground engineers from the cockpit window and spoke to them on the intercom. It did not occur to me that the problem was of my making and when the truth was known I had to ask for instruction on how to unpressurise the aircraft in order to open the door.

The Tudor was never an aircraft which many of us in BSAA could honestly claim to like. In theory the journey to Bermuda should have been shorter and smoother. The improved true air speed at the higher level at which the flight was planned should have ensured this result. Certainly the pressurised cabin enabled flights to be conducted above ordinary alto-cumulus cloud but the cabin heater failed so frequently that it was necessary to descend in order to avoid freezing everyone on board. This made navigation difficult because at 10,000 feet the moon, planets and stars were often invisible for hours at a time. No astrodome had been provided for star shots because experience on the Constellation had shown that the glass could fracture under the differential pressure and doom the unfortunate navigator to be sucked out for a grisly death. The flat glass substitute was far inferior.

Purely from the pilot's point of view a tricycle undercarriage would have been vastly preferable to the tailwheel arrangement. The Tudor had a very large tail surface and it was difficult to keep the aircraft straight when taxying on icy taxiways in blustery conditions. It was also unfortunate that the range of the Tudor, sixteen hours flying, was less than that of the Lancastrian. To remain within the maximum permitted all-up weight the fuel tanks could seldom be filled to capacity unless the load was reduced by removing cargo. When this had been discharged in the Azores and the captain operating the following service had for his own good reasons been unable to accept it there had been well-publicised blasts of outrage from the shippers. On the Lancastrians Captains had learnt not to trust the forecast winds, but however long the flight sufficient fuel had always been carried. As there was no airport within range of our aircraft if something prevented an immediate landing Captains carried at least two hours fuel for holding over Bermuda beyond the requirements for the flight itself.

On this occasion the flight plan envisaged a journey lasting twelve hours at 25,000 feet. The first position fix which I obtained showed the wind to be stronger than expected and within another twenty minutes the heater failed. Cursing his luck the Captain descended into the heavy cloud and rain and levelled out at 10,000 feet. When I was next able to get a fix we were forty miles north of track and our expected time of arrival at Bermuda had to be put back by over an hour. We had been flying for ten hours when I was able to plot another fix and there was no expectation that we could reach Bermuda in less than another three and a half hours. When we had been in the air for thirteen hours we were all staring anxiously through the cockpit windows. We were in VHF contact and the radio officer was obtaining good bearings but we were not confident of the distance we still had to fly.

"Drinks on me to the first man to see the lights of Bermuda," the captain muttered.

A few minutes later lights were seen and an exhausting flight concluded, thirteen hours and forty minutes after leaving Santa Maria. On my next trip the sector to Bermuda took just over fourteen hours. There was an alarming occasion when one of the Tudor captains declared an emergency and a USAF search and rescue aircraft flew out to accompany the airliner with every expectation that it would ditch in the sea short of its goal.

After this incident had prompted the Governor of the colony to make representations to Whitehall, Captains were advised to consider the option of routeing via Keflavik in Iceland or Gander. Readers whose vision of the world has been obtained from the Mercator projection should look at a globe or conical map to appreciate the reasons. But given the severe weather over the northern Atlantic in the winter neither were ideal staging ports. Even so both were used. One of our Captains attempting to land at Gander in poor weather was offered a ground-controlled approach by the controller. He declined because he was sure that Bennett would expect him to be able to land without assistance and, just as important, would deplore the cost involved as a charge was made in dollars for the service.

When several other pilots who were stacked above the Tudor became impatient and abusive the controller said: "You had better have this one on us."

At the end of 1947 BSAA suffered the destruction of two Lancastrians, both on the airfield at Bermuda. On the first occasion the aircraft arrived as a violent thunderstorm was lashing the island with torrential rain and the wind was gusting strongly. Much of the runway was flooded and the airfield's lighting had failed. The Captain circled for an hour hoping for an improvement before making his approach. The undercarriage collapsed on striking the flood water, but nobody was hurt. The other Lancastrian had left Bermuda for Heathrow when the Captain had to shut an outboard engine down. To reduce weight to an acceptable level he attempted to jettison fuel but the system malfunctioned and the crew were sickened by petrol fumes. A heavy landing damaged the aircraft beyond repair but there were no serious casualties.

Shortly afterwards Bennett arrived in Bermuda, having negotiated a price with the insurers for the Lancastrians which had been written off. The cylinder blocks on the engines and other items were serviceable and valuable as spares. To assist the station engineer he mustered all the crew members including the stargirls who were on the island and put them to work removing the parts which he wanted.

More bizarre than these events was a spate of stowaway incidents. In the first a man entered a Tudor at Heathrow when it was parked outside a hangar awaiting movement to the terminal building. He found a trap door under the carpet in the passenger cabin and concealed himself under the floor. Once the aircraft reached its cruising level the intense cold proved too much for him and the passengers were astonished to see the carpet bulging upwards as he emerged.

Another case involved a Portuguese youth whom nobody noticed as he climbed up the undercarriage leg of a York at Lisbon airport into the engine nacelle. He succeeded in keeping his body clear of the space into which the wheel retracted after the aircraft had taken off for Heathrow. He survived the cold, noise and vibration but was spotted by an airport policeman when he was trying to make his escape.

The third incident was followed by the prosecution of the stowaway for endangering the lives of the crew of a Lancaster freighter on which he had concealed himself. En route to Bermuda he had startled the crew by appearing suddenly in the cockpit. To everyone's surprise his defending counsel proved to the satisfaction of the court that his hiding place near the tail section could not have moved the aircraft's centre of gravity out of the safe range and he was acquitted.

Early in 1948 I was promoted to Captain and moved from the flat near Uxbridge to the house which I had bought in Chalfont St Peter. I had obtained a loan against the security of my life insurance policy. In those days pilots and others who flew for a living had their policy premiums loaded. After putting down the deposit on the house I had just enough money left to buy a ten-year-old car from another pilot who had urgent need of funds. The car had a canvas roof and no sidescreens and as soon as I had covered five miles a warning light accompanied a disturbing rattle which indicated that the lack of oil pressure was certain to lead to the failure of the 'big end.'

I had not counted upon being required to pass a driving test. These had been discontinued on the outbreak of war in 1939 and had only recently been reintroduced. I took the test in Slough on a bitterly cold morning. The examiner looked at my car with unconcealed distaste.

"That *and* excruciating toothache!" he grumbled.

I was unaware that I had made any unpardonable mistakes but when the test concluded the examiner produced a printed pad and began underlining some of the sentences.

"I don't know where you have been driving," he said, "but your standard is well below what we tolerate here."

I read the underlined passages.

"What's this about lack of consideration for other users of the road?" I protested.

"I don't have to explain anything. Those are the parts of the test which caused you to fail. However when I told you to turn left at the traffic lights the cyclist on your near side almost fell off to avoid being struck by your car."

"Hang on," I expostulated. "I got to the lights first and stopped because they were at red. That cyclist crept up on my left when I was stationary. He should have given way to me."

The examiner sneezed and a paroxysm of pain made him put his hands to his face. "Good morning, Mr Jackson, and please read the highway code before you apply again."

Then he extricated himself from my car and hurried away to his office. After that I had some lessons from a driving school and passed the test a month later. Further embarrassment followed my discovery that the car's fuel contents indicator was unreliable and as a precaution I kept a gallon can of petrol in the boot. Inevitably the tank ran dry when I was in uniform and the car glided to a halt by a bus queue. It was impossible not to overhear the caustic contents as I funneled in my emergency supply.

"He won't be able to do that in the air, will he?" one wit remarked, and worse still: "If I knew what airline he works for I should avoid it like the plague."

I enjoyed being in command of an aircraft again, even the slow old York and grew accustomed to the questions nervous passengers used to ask.

"Can this aircraft continue to fly if three of the four engines fail?"

Captains were usually advised if anyone of importance was on board. As I was signing the load sheet the traffic officer at Heathrow told me that a Scotland Yard officer was accompanying a wanted criminal to face trial in West Africa and pointed out their names on the passenger list. Some time later and before I had gone into the cabin I asked the stargirl where the couple were seated, mentioning the relationship.

She was most surprised. "Someone was pulling your leg. Those two are certainly well acquainted. Between them they have drunk most of our stock of beer."

Peering through the curtain I saw the couple laughing uproariously and when I reached them after chatting to others first I had discounted the story which I had been told. Even when after a few minutes conversation one of the men revealed that he was a police inspector I did not believe that his companion was a crook. Only after an interesting discussion on the sensational acquittal of a man who had almost certainly murdered a young woman in Regents Park did I notice that the other passenger had remained silent and was looking very forlorn. Some weeks later I read of the conviction in Sierra Leone of a former hospital superintendent who had fraudulently drawn the salaries of non-existent employees whom he had pretended were on the payroll.

More interesting were a group of experienced flying boat pilots whose brief it was to survey the South American route for suitable landing sites for the Saunders Roe Princess which Bennett was keen to buy. One of the pilots was Geoffrey Tyson whose autograph I had obtained in 1935 at an air show in Littlehampton when he was employed as a stunt pilot by Sir Alan Cobham.

Bennett also intended to buy the de Havilland Comet but for the present we needed sufficient Tudors to replace the Yorks and Lancastrians. Air France had already put the Lockheed Constellation into service and had plans to introduce the larger Super Constellation which would operate Dakar to Rio de Janeiro without a refuelling stop en route. When we were not flying Yorks to South America we would be practising landings on the Tudors made available to us through BOAC's refusal to operate them. These included the original Tudor I which was also unsuitable for our routes and same Tudor IVs which had been fitted with a flight engineer's station.

In January 1948 BSAA suffered the first of the two disasters which were finally to lead to the company's extinction. The Tudor *"Star Tiger"* left London to fly to Bermuda via Lisbon and the Azores. On the first sector the cabin heater failed, a compass malfunctioned and there was an engine problem. The following day the flight was resumed to the Azores but the flight plan time to Bermuda was so long that the captain decided to remain at Santa Maria overnight. When the Tudor took off it was with full tanks and in fact overloaded by 1,000 lbs. The flight plan anticipated a time of twelve and a half hours. Three and a half hours later Bermuda was advised that the arrival would be an hour later than originally estimated. For some hours the navigator had no opportunity to obtain a fix but when the Tudor had been airborne for eleven and a half hours Bermuda was informed that it still had 550 miles to go. On two occasions *"Star Tiger"* obtained a bearing from the air station then no further contact was made.

Without much delay the alarm was raised and a number of aircraft began a search but no wreckage was found. The search aircraft included a Lancastrian which had left the Azores about an hour ahead of the Tudor and had been in radio contact with the crew en route. The Captain stated that on the last stage of the flight there was no icing, turbulence or fog. An examination of all the Tudors and their components could find nothing to account for the sudden catastrophe. The Air Registration Board advised Lord Nathan, the Minister for Civil Aviation, that it had no evidence to justify grounding the Tudors but the latter decided to do so. Bennett was so outraged that his own authority as Chief Executive of the airline had been overridden that he gave an interview to the *"Daily Express"* complaining of interference by those whom he described as "totally ignorant of aviation" and pulled no punches in criticising the Government's interference in matters outside their competence. He had not consulted the chairman of BSAA nor any member of the Board before taking this initiative and his refusal to retract his statement prompted a request for his resignation. When he refused to accede to this he was dismissed.

By a strange irony his replacement was Air Commodore Brackley, deputy to the chairman of BOAC. In 1935 as Major Brackley he had been Air Superintendent of Imperial Airways when

Bennett had joined that airline. I was introduced to the new Chief Executive at the Company's headquarters in King Street, off St James.

"I hope I will fly with you before long," he remarked, and added: "I have been in the aviation business so long that just sitting at the back among the passengers I can tell at once whether the pilot is any good."

Brackley also met a delegation of three stargirls who were most displeased that stewards were also to be employed in addition to young women. Their anxiety centred partly on the question of seniority. They were reluctant to take orders from the newcomers for no other reason than that they were men. In the short time that Brackley was with us that question was never resolved.

The company continued to have accidents. I was in Santiago when I received a telephone call from our office notifying me that my crew would have to remain an extra day or two. A York on its way south had incurred an uncontrollable fire in an engine and crashed trying to land at a small airfield in Brazil. There were several fatalities among the passengers. The Captain had been involved in one of the crashes on Bermuda's airport. After this he did not remain with us. Airlines hate to employ an unlucky pilot in case the travelling public remember his name. Perhaps the most notable example of this evident truth was the unfortunate Captain of the BEA aircraft which crashed in Germany with the Manchester United football team on board. A long campaign by the pilots' association which finally proved that the cause of the crash was slush on the runway did nothing to repair the ruin to the Captain's career.

Another Lancastrian had an engine failure on take-off from Lima. One undercarriage wheel struck a wall as the pilot tried to climb. Subsequently he was able to make a smooth belly-landing and nobody suffered a scratch. While he had been circling to use up fuel and reduce weight all of Peru were kept up to date by radio reports and hundreds of cars drove to the airport to watch the outcome. The President of Peru was so impressed that he had the Captain driven round the city in an open car and later a gold medal was presented to him.

Bill Flower, one of our Captains, confided to me that his wife had psychic powers. She always knew when she was passing close to a person on the point of death even if the individual was out of sight in an adjoining house. Walking in company with her husband past a line of parked airliners on the north side of Heathrow where the terminal used to be she had clutched his arm.

"That aircraft," she indicated with a gesture, "is going to have a terrible accident very soon."

It was a Pan American Airways Constellation which a few hours later hit a low wall on the approach to Shannon airport. There was only one survivor.

About a year after the appointment of Brackley Flower paid a short visit to the head office in King Street and left his wife to wait in the entrance hall. When he reappeared she was in an agitated state.

"A man has just left the building," she confided to her husband, "who will lose his life in a few days time."

Flower went up to the hall porter and asked if he knew who had just gone past his cubbyhole into the street.

"Of course I do!" he was told. "Air Commodore Brackley just went out."

The following day the chief executive flew out to inspect BSAA's stations on our routes. Arriving at Rio de Janeiro he found time for a swim on Copacabana beach. The sea was dangerous on that day and he was drowned. His body was recovered and flown to England and I was one of the Captains detailed to attend his funeral in London.

The final chapter of the BSAA story was about to be concluded. In January 1949 the Tudor *"Star Ariel"* took off from Bermuda for Jamaica and an hour later it was cruising in good weather at 18,000 feet when it transmitted a routine position report. This was the last message to be received and as in the case of *"Star Tiger"* no distress call had been broadcast. A thorough search revealed no wreckage. As this was the second unexplained loss within a year the certificate of airworthiness of the Tudor for passenger carriage was revoked. Having regard to the fact that the airline was largely dependent upon its fleet of Tudors it could no longer continue operations and the South American route was transferred to BOAC.

Reflecting on the demise of the company nearly fifty years after these events it may surprise the reader to know that we were a very loyal and happy group of employees, very responsive to Bennett's anxiety to avoid overmanning and bureaucratic wastefulness and just as keen as he was to make a success of the enterprise. We were aware that we did not operate aircraft which were adequate for the job but believed that we had suffered an amazing run of bad luck. It was also true that the flying experience of most of us had been gained in wartime conditions when the difficulties imposed by weather and technical deficiencies were a challenge to be overcome by courage and effort. It has to be acknowledged that paying passengers were entitled to expect caution and safety to have been the paramount considerations and too often they were not.

When *"Star Ariel"* was lost the western allies were engaged in supplying Berlin by air in face of a land blockade by the Russians designed to test the will of their former allies to preserve their unity at the risk of armed conflict. Bennett had formed his own private company using Tudors and flew them as oil tankers. Surplus Tudors were also being operated on the airlift by BSAA crews based in Germany. Unheated, unpressurised and without provision for passengers these aircraft continued to supply Berlin until the Russians acknowledged defeat and raised the land blockade. Thereafter, following modifications to the Tudor, Bennett eventually was able to regain a full certificate of airworthiness permitting passengers to be carried. Until the day of his death in 1986 he remained convinced that sabotage had been the cause of the sudden disasters over the western Atlantic Ocean.

CHAPTER 11

THE SPEEDBIRDS

When I had visited Heathrow for the first time early in 1947 the airline companies were using as their offices marquees, portable sheds and former military mobile wagons. These were all adjacent to the Bath Road on the north side of the airport, as was the control tower and the aircraft parking area. Three years later little had changed. The marquees and wagons had gone but in their place was a mass of prefabricated huts. There was nothing to indicate that the terminal would be resited in the centre of the airport. There was discussion about the feasability of extending the London Underground railway to the airport from the station at Hounslow West. Another thirty years were to pass before this happened.

The flight crews of BSAA were generously treated when we were absorbed into BOAC with our existing ranks and seniority, particularly as the senior corporation already employed far too many people at all levels. The new chairman Sir Miles Thomas, whose huge office in London overlooked the Ritz Hotel and Green Park, was astonished at the atmosphere of luxury so totally out of accord with the dreadful financial results of the airline. He had not only to reduce the number of staff and close down half a dozen aircraft bases but also to dispose of many of the uncompetitive types of airliner in service.

Among the latter were the flying boats. There had been several derivations of the Empire class which had first plied the routes in 1936. The Sandringham appeared in 1947, the Solent a year later. The passengers enjoyed travelling in them, particularly those who spent four or five nights ashore en route to South Africa; but the aircraft were so slow that their continued use on international routes made economic nonsense. By 1950 the decision was taken to rely upon landplanes and to close down the flying boat base at Southampton. To replace the Yorks and Lancastrians BOAC ordered twenty Handley Page Hermes airliners and twenty two DC-4M Argonauts. The former, a pressurised airliner carrying about thirty five passengers, was the first British aircraft of that size to be provided with a tricycle undercarriage rather than a tailwheel. Like the Tudor it had suffered a disastrous inception when a prototype crashed due to the reversal of one of the control systems during servicing. The economics of the Hermes in no way matched those of the competition but it remained in service for a few years.

Most of the pilots of BSAA were assigned to fly the Argonaut, a pressurised version of the Douglas Skymaster but powered with Rolls Royce Merlin engines and built under licence in Canada. The manufacturer had agreed to allow BOAC to pay for them concurrently with the revenues they earned. A smaller number of pilots began training on Lockheed Constellations. The British Government had felt obliged to allocate scarce dollars to buy these in order that BOAC could offer a service on the North Atlantic. During the war the priority had been to build bombers and fighters and no British airliner had been developed which could cross the North Atlantic with any sort of payload.

We were given a course of ground study lasting several weeks before we sat the technical examination on the Argonaut. This was a change from the BSAA system of providing some notes and leaving us to study in our own time. At any rate we all passed and were then given lectures on survival after a forced landing in a remote desert or at sea. This involved an understanding of the purpose and use of the drugs in the aircraft medical kits. All crew members were regularly briefed in these matters and those new stewardesses who had been nurses were initially shocked that non-medical staff were told how to inject drugs which in normal practice were only used under a doctor's supervision.

The effect of lack of oxygen (anoxia) was demonstrated in an oxygen chamber. Four of us sat playing cards as the oxygen level was reduced to simulate the equivalent decrease at ascending altitudes. Two who had been told not to wear masks eventually ceased to respond to speech or gesture. Connected to oxygen they revived almost immediately, unaware of any interruption to the game. The other couple then had their oxygen supply disconnected and quickly lost consciousness. I had been a subject of this quite painless practice when in the RAF, which had no pressurised aircraft but necessarily had to provide oxygen to many types of machine. During my time in BOAC a steward died in the oxygen chamber, presumably as a result of some physical condition unknown to him, and these demonstrations were discontinued.

Flight training on the Argonaut was practised at Heathrow. Although nothing like as busy as it was to become I recall one night's 'circuits and bumps' when the tower controller told us that an increasing number of complaints about the noise we were making were being passed to him. I also remember our instructor showing us how best to land an Argonaut when two engines on the same side were shut down. On the final approach he closed the throttles of the two engines and flew towards the runway allowing the aircraft to descend no lower than about six hundred feet. When the runway threshold was no longer visible from the cockpit he extended the flaps to their full extent, closed the other two throttles and allowed the aircraft to descend steeply before rolling out and touching down. It worked very well and we stopped within the runway length but I wondered what those who were watching this unusual manoeuvre imagined the pilot was trying to do. Before long the airlines were ordered to discontinue flying training at Heathrow and when simulators were introduced in the later 1950s 'recovery from unusual attitudes' was practised on these.

I had hoped to return to the South American route after finishing the conversion course and was very disappointed when I was told that sufficient pilots had already been selected. I was not particularly surprised. I encountered a Canadian pilot whom I remembered from the Empire boat days of 1942. He had been in BOAC service since before the war. I recalled that he had then been trying to get a posting to the corporation's Atlantic ferry operation centred on Montreal, where his wife lived. He was unsuccessful. One of the BSAA pilots had previously flown Constellations for KLM. When we were absorbed into BOAC he expected that this would be reflected in his posting but he was put on the Argonaut course. The manager of the Argonaut fleet was Jack Harrington, now deskbound, and I managed to see him in his office. He remembered me and showed a friendly understanding but explained his situation.

"I employ dogs," he remarked, referring to his deputy, who was another former captain, to his Flight Superintendent and two Flight Captains. "I expect them to do my barking for me."

My initial Argonaut flight with passengers was therefore east-bound to Calcutta. The callsign for all BOAC aircraft was "Speedbird" and the first transit stop was Frankfurt where BOAC aircraft were handled by British European Airways. On the way from Heathrow on a clear sunny afternoon I noticed that my co-pilot was carefully following our progress on a topographical map although we were navigating from one radio beacon directly towards another.

"There should be a substantial crater somewhere here," he told me, pointing at Koblenz on his map. "My Halifax went down in flames before we had reached our target and dropped the bombs. This is my first trip back to Germany since the end of the war."

At Frankfurt airport a BEA traffic officer showed me the way to the weather office and to air traffic control where I filed a flight plan to Cairo. For the next couple of years he was invariably the one who performed this duty. I thought I detected the trace of a Welsh accent. One day I asked him why he had remained so long on the station. Ground staff were usually moved from one place to another quite often.

"I live here," he said.

"Do you mean you married a German girl?" I asked.

"I am a German."

He laughed, seemingly pleased that his English was so good that I had been deceived.

"I was a prisoner of war for three years and worked on a farm in Wales."

From time to time I stayed overnight in Frankfurt or in Dusseldorf and knowing how savagely both had been bombed during the war I was astonished at the speed with which the rubble had been cleared, the cities rebuilt. London bore its scars for very much longer.

We landed at Cairo during the hours of darkness and as the aircraft doors opened the distinctive smell of the place, like spilt petrol, reminded me of previous visits during the war. In the Customs hall an official looked carefully through the pages of English newspapers which someone had retained.

"He is looking for cartoons of King Farouk," it was explained to me. "They have been told to confiscate any newspapers showing disrespect to that fat old pervert."

The traffic officer asked me if I wanted to stay at the Heliopolis Hotel or to go along with the rest of my crew to one in the centre of Cairo. Unhesitatingly I chose the second option but I soon discovered from the comments of other crews residing in this hotel that my choice was unwelcome to some of them.

"We have been complaining to BOAC that this place is not good enough," I was told. "During the hot weather the atmosphere is horrendous and the food is poor. Then the management show us a list of captains who have chosen to come here rather than stay at the Heliopolis Palace and we can make no progress at all."

I apologised for my ignorance and my informant nodded sadly.

"So you are one of the BSAA lot," he remarked. "I had hopes of a command within a year or so but not any longer. The old sweats of Imperial Airways show no inclination to retire and now you people come in through the back door."

Many of us were rather sensitive to observations which we were probably too inclined to regard as slights. I came across a Captain with whom I had flown on the Empire boats and reminded him of that fact, adding that I had more recently been flying with the defunct airline.

"That inferior mob!" he replied. "Poor you."

As time went by and we found ourselves accepted in the most friendly way we settled down with our new colleagues very well. We discovered which very senior Captains preferred to be lodged in hotels apart from their crews and became accustomed to the habits of former flying boat Captains who signed reports and added the word "Master," used words like 'forrad' and 'astern' and called a door a 'hatch.'

Flying east the journey took us to Bahrein in the Persian Gulf. This had been a staging post for both landplanes and flying boats in the era of Imperial Airways and there was a BOAC resthouse there. In 1950 only Bahrein and Kuwait were served by four-engined landplanes and the latter had no buildings nor runway. A windsock indicated wind direction and some black oil markings on the hard sand revealed the area on which it was safe to land. Two small tents were used to handle the passengers and the motor cars which carried them to and from the airfield sent up clouds of dust. So too did the hooves of horses ridden by Arab traders who came to meet incoming aircraft with their own mounted escort of armed men to protect their gold and silver.

Today Kuwait and Bahrein have been transformed into modern cities with large hotels to accommodate visiting businessmen. So have Dubai and Abu Dhabi whilst Muscat has also emerged from the Middle Ages. Yet even in that past era of aviation the Sheikhs were keen air travellers.

A former BSAA radio officer, new to the region, was distracted by the attempts of an Arab passenger to enter the flight deck. After slamming the door in his face several times he asked the co-pilot to ring the call bell for the steward.

"There's some old wog trying to get in here," he complained when the steward appeared. "If he wants the toilet show him where it is but keep him out of here."

The steward withdrew but was soon back.

"The Sheikh is anxious to hand out some gifts to you," he revealed. "Do you still want me to keep him out?"

The rulers of the Trucial states as they were called in those days were traditionally generous and often presented gold watches and silver cigarette cases to the pilots with whom they flew. The gifts were sometimes carried in old Huntley and Palmer's biscuit tins fastened by a Yale lock. Inevitably as time went by and perhaps because Allah was generally merciful to air travellers the practice ceased. In subsequent years oil revenues transformed the economies of the local rulers whose despotism has extended from suzerainty over their own subjects to autocratic instructions to the airlines whose aircraft come and go through their airports. If a ruler has a sudden whim to travel with five of his wives and fifteen other members of his family it would be a bold airline manager who regretted that he could not accommodate them, the First Class Cabin being already fully booked. I have known a Chairman of BOAC to alter the published schedule of a VC-10 service so that a ruler could have the use of the aircraft for a few hours.

Landing at Calcutta's Dum Dum aerodrome where I had been stationed with 52 squadron six years earlier only the aircraft standing upon the tarmac looked different. The road into the town with its potholes was still awful, progress interrupted by water buffalo drawing carts, sacred cows and many jaywalkers. The airline bus on which I travelled contained both my crew and our passengers and I entered into conversation with one of these.

"Did you make any use of the aircraft lounge?" I asked, referring to the horseshoe arrangement of six seats in the tailplane area of the Argonaut.

He shook his head vigorously.

"Never went near it. I stayed in my seat near the galley so that I would be among the first to hear of any emergencies. I only went to the lavatory when it was imperative to do so."

He went on to tell me of his terror of flying and his attempts to persuade his employers to send him to Calcutta by sea. He had still not composed himself after a totally uneventful trip and was smoking one cigarette after another.

"You have arrived safely here now," I reminded him. "I don't suppose you will have to fly again for some time."

"Oh! I will, I will," he replied, lighting another cigarette.

"What do you do?" I asked.

"I am a deep sea diver. I get sent all over the world."

In the immediate aftermath of the war as small private airlines with inadequate financial resources bought surplus Dakotas and converted military aircraft and tried to make some money there were a great number of accidents. By 1950 pilots' instrument ratings had been introduced, to be renewed by examination each year. Check lists for each aircraft were published so that the flight crews did not have to rely upon their memory to perform all the necessary actions before taking off or preparing to land. Resistance to some of these measures sometimes came from those who prided themselves on their superior competence.

"Surely you aren't about to take off with your landing lights on?" a training captain challenged me.

During the war we had been trained to take off and land without using these in order to avoid attracting the attention of enemy intruders. To deny oneself in peacetime was no more than bravado.

Taking off at night the landing lights were useful in scaring off the runway stray animals such as dogs or even larger animals. One of BOAC's instructors who had been sent to the government centre to qualify as an accredited instrument rating examiner told me that he incurred derision when he was discovered to have acquired a check list for the small aircraft he would have to fly, one on which he had little previous experience. There were some Captains who would forego a scheduled night stop in order to make up a lengthy delay to a service even if this action involved over twenty consecutive hours on duty. A few more years were to pass before legislation was passed to prevent this happening.

Until the end of 1951 I flew regularly through the Mediterranean and the Near East to Calcutta. There was a pleasant halt along the way when a rest was taken in Rome on the homeward journey. A number of BOAC services passed through each day and a crew change was always welcomed. Evidently United States air force crews also shared this view because I remember noticing a group of officers headed by a colonel assembling in the lobby of our hotel.

"On the next leg," the latter informed his companions, "*I* shall navigate."

"Certainly sir," a major replied. "We have a sextant on board."

"How's that again, Ozark?" his superior enquired, his brow wrinkling.

"A sextant, colonel: for taking star shots."

"A sextant!" The would-be navigator shook his head. "I wouldn't know one from a birdcage. I will only navigate if I can see the ground below the airplane."

More often than in Rome our breaks were in places like Cairo, Damascus, Teheran or in the Gulf. In Karachi one day I encountered Air Vice Marshal Bennett with the crew of his Tudor. He and I were both bound for Bahrein and the following day I had just taken off when I heard him asking for permission to taxy out to the runway.

"Negative," the controller replied. "Your documents are not in order. We are advised that your cargo includes crates of tea which are not listed on the manifest."

"Nonsense!" Bennett replied. "I have not taken on tea. Give me my clearance at once. If you remove any of my cargo I will make you liable for demurrage."

For as long as I remained within range I listened to a fascinating charge and countercharge. I did not learn the outcome until a few days later when I saw Bennett again.

"What was all that about demurrage?" I asked him. "Come to that, what *is* demurrage?"

"Compensation for an unwarranted delay," he informed me. "I hadn't accepted any consignment of tea."

"What gave them the idea that you had?"

"Well, I think one of my crew had bought some for his own purposes."

One of the differences of flying over the Indian subcontinent in the post-war era was the administration of air traffic control. Inevitably this was no longer conducted by the British and stories were gleefully related of the consequences. A pilot asked the airport controller for the surface wind.

"There is a sideways wind."

Another pilot was ordered to hold over a radio beacon at 4,000 feet. A different pilot promptly broke in: "No, no, you gave *me* instructions to maintain 4,000 feet over that beacon."

A third pilot added his voice. "I have been circling that beacon at 4,000 feet for fifteen minutes."

To which the controller replied: "Oh! my God! What to do?"

To be fair the new men learnt their trade very quickly as did the locally employed mechanics who serviced our aircraft. They worked under a station engineer sent out from England.

On one of my trips I came across George whose involvement in the perfume trade between West Africa and Rio de Janeiro had led to his departure from amongst us. He was still up to his old tricks and with a broad smile told me that there were far more opportunities for his entrepreneurial talents east of the Greenwich meridian. He came to grief once more however when a wrapped package which he had asked an unsuspecting crew member to leave for him in the care of the Karachi resthouse was dropped on the aircraft steps and burst open to scatter to the winds a not inconsiderable quantity of currency notes. George left the employ of BOAC but found work with another airline and I was to meet him again on a future occasion.

In December 1951 I was delighted to return to the South American route. I knew those crews who had been flying across the South Atlantic and became acquainted with the non-pilot navigators whom BOAC had assigned to operate among them. To keep my own navigation licence valid I resumed my custom of navigating the oceanic sector eastbound or westbound myself.

The Argonaut carried one steward and one stewardess and those former stargirls who spoke Spanish or Portuguese were still flying to South America. Among the large number of stewards whom

BOAC employed there were inevitably a few homosexuals, some less discreet about their inclinations than their brethren.

One Captain's voyage report contained the robust statement: "I am disgusted at having to fly in company with queers."

He was told that he was shouting into the wind. The stewardesses found it less of a problem, "rather like having one's sister on board" said one. Passengers appreciated their meticulous attention to cleanliness of the washrooms, regularly re-supplied with soap and towels.

I recall one steward who could keep a group entertained for hours with his anecdotes, gales of laughter attracting others towards the gathering. Someone mischievously asked him how he had enjoyed a trip he had made with one of BOAC's most attractive stewardesses. He rolled his eyes.

"Oh! what a lovely girl – almost made me turn normal."

Another steward on my crew spent the long night over the Atlantic seated in the galley typing short stories. They had unusual titles like "The man who fell up." He was particularly put out by the habit of South American passengers of ignoring the illuminated signs to fasten their seat belts and walking about at those times when he and the stewardess were trying to serve meals.

"I can't do a thing with them," he complained to me, wringing his hands. "I have begged, I have pleaded . . . ". His voice rose into a shrill crescendo: "I have been in tears and still they won't sit down."

There was only a twenty four hour stopover in Santiago but that was just sufficient time for me to visit my parents who lived on the coast at Vina del Mar. My younger brother and sister also lived with them. I returned to Santiago by train on the following morning. To accommodate the crews BOAC had built a comfortable resthouse in the outer suburbs of the capital in addition to a house for the station manager. The reason for this was the refusal of the Bank of Chile to allow revenue earned to be changed into sterling and repatriated. A Constellation crew also arrived from London having flown via the Caribbean but the distance involved and the number of stops neant that bad weather or a mechanical delay played havoc with the schedule. The solitary weekly flight was so seldom on time that even the most ardent Anglophile resident in the continent preferred to choose another airline operating at greater frequencies even if it involved changing aircraft once or twice along the way. So BOAC abandoned the west coast route.

I was contentedly flying to and from Santiago for a couple of years suffering only from irritating misunderstandings with one of the "dogs" on whom Captain Harrington depended to do his "barking" for him. Misunderstanding is the word I use in the mellow spirit induced by the passage of about forty years although at the time I became exasperated. The Deputy Manager was a dismal little bureaucrat who looked for opportunities to exert some authority.

Captains' salaries were raised annually by fifty pounds for four years at which point the top of the scale was reached. The D.M. sent for me one day and invited me to take a seat. Then he rang for his secretary.

"Bring me Captain Jackson's incremental slip," he ordered her.

She returned with a minute piece of paper about the size of a receipt for a box of matches. He took this from her and presented it to me.

He could see that I was perplexed by such a performance over a straightforward accounting exercise.

"You fellows from BSAA seem to think that this rise is automatic," he said. "It is not. It is only given if I receive reports from the Flight Superintendent which show consistently good work on your part."

I did not quarrel with him as I had received the increment but shortly thereafter was told by another BSAA Captain that he had not had his salary raised. Nor had he been able to discover in what respect he had failed to perform his duties.

The next occasion on which I had an encounter with the D.M. was when he wished to know why I had signed for a cheese sandwich at Rome airport when there was a contract in force for a crew meal to be supplied there. The explanation revealed something of the odd machinations within BOAC. Although the airline's navigators were perfectly capable of preparing a flight plan a host of operations officers had been specially trained and installed at stations down the line. One of their duties was to have a flight plan ready for the following sector of every aircraft in transit.

The various aircraft fleets were separately audited and Captain Harrington was indignant that his was being debited for work which navigators had always been required to do. He hoped to have the system changed by refusing to allow his Captains to accept the prepared plans. So when my crew were having a meal in the airport restaurant the navigator was hard at work on the flight plan, taking occasional bites at the sandwich which he had ordered and I had signed for.

"These bills add up to a considerable sum over a period of time," the D.M. protested.

Another Captain told me how he had driven all the way from his home in Bournemouth on a summons "to discuss an important matter."

"I had written a voyage report about an incorrectly fitted propeller," he told me. "It could have come off and caused an accident. I assumed that we would be discussing that. I was wrong. He told me that he had passed my report on to the engineering department. He wanted to complain about unauthorised consumption of sandwiches during transit stops."

A more serious principle was involved on the next occasion that I had upset him. I was in Dakar and due to fly to Rio de Janeiro with a refuelling stop at Recife, a journey lasting nearly fifteen hours from take-off to the final landing. The Argonaut arrived at Dakar with a technical fault which eventually took about five hours to correct, all of which time we were on the airport. Consequently I did not hesitate to signal to Recife that on arrival I would be remaining a sufficient time for the crew to rest before continuing to Rio. This brought a written rebuke from the D.M. As no legislation had yet been introduced to define acceptable periods of duty there was nothing I could say beyond emphasizing that it had always been the Captain's prerogative to decide whether to continue a flight in such circumstances.

Shortly thereafter I received word from a sympathetic Flight Captain that I was not the D.M.'s favourite pilot and before long we met again. I had flown from Santiago to Buenos Aires and had still to continue to Montevideo and Rio de Janeiro. On the way to the Argentine capital I had observed on the southern skyline the approach of an extremely active front with towering thunderclouds and vivid flashes of lightning. A front of that intensity is called a 'pampero.' In the weather office on the airport the forecaster's estimate of its progress indicated that it would probably pass north through Montevideo when I was refuelling there, a very short distance across the River Plate which separates Argentina from Uruguay. As it happened a full complement of passengers was being embarked in Buenos Aires and no one was destined for Montevideo. With the agreement of the station manager I decided to take on sufficient fuel and proceed directly to Rio de Janeiro without an intermediate landing. As a result a very rough flight was avoided.

On my return to England the D.M. sent for me and demanded an explanation which he could perfectly well have found in my voyage report. He did not accept it and muttered about mail bags which had been awaiting collection in Montevideo. Then he opened a thick tome of BOAC regulations and read one of them out to me in triumph.

"'Scheduled stops shall not be overflown.'"

He went on to say that the Uruguayan airport authorities were deeply offended and that my reappearance there would be an embarrassment to the airline. I would be flying on routes other than to South America in future.

I got my own back in a childish tit-for-tat way a few months later. I was flying from Cairo to Rome where another crew were scheduled to take over my aircraft. Arriving overhead I found the airport in the midst of a snow storm with gale-force winds. I diverted to Nice and was followed in there by a KLM aircraft. The Dutch Captain asked me what I was going to do and I told him that I was going to head back for Rome when the storm had abated.

"Fine," he said. "I will go on to Amsterdam so I will have my Rome passengers transferred to your aircraft."

We had arranged this and I was about to leave for Rome when I received a message from London authorising me to fly on to Frankfurt and then complete the journey to London.

"Negative," I signalled back. "I have twenty five passengers for Rome. Scheduled stops shall not be overflown."

I don't recall any further unpleasantness between us but I was aware that other Captains had similar experiences. Perhaps the attitude of the D.M. was attributable to the improvement in salaries paid to the pilots. Equally welcome to us was the payment of meal allowances which permitted us to eat where and when we liked and not go hungry because we were asleep or on an airport bus when meals in the hotel were being served. His own flying had been in the bad old days when as in the BBC the glory of the organisation was supposed to justify the ungenerous salary scale.

Captain Harrington once said to me: "If you pilots get another pay increase I shall ask to be relieved of this post to go back to flying."

The introduction of Tourist Class in 1952 saw the seating in an Argonaut increased to accommodate fifty five passengers in narrower and more closely confined rows. There had been no great luxury in the previous one class configuration and initially the food offered to the tourists was cold and rather meagre. Holidaymakers may have welcomed the reduced fares but the great number who were travelling at the expense of their employers were disgruntled to say the least. This was particularly the case on the West African routes where before BOAC had introduced daily services to Ghana and Nigeria the expatriate residents had always enjoyed the voyage to and from Britain by sea. The time spent on the comfortable ships of the Elder Dempster Line did not count as part of their leave.

I remained on Argonauts until 1958 by which time two had been destroyed in Africa. In both cases weather conditions played an important part. Tripoli airport in Libya was the former Castel Benito airfield built for the Italian air force. It had never been improved and boasted one radio beacon. It was shrouded in mist when a pilot northbound from Nigeria tried unsuccessfully to land. If there is fog or very low cloud in that area the likelihood is that it will affect the alternate airport on Malta also. The following year an Argonaut crashed at Kano in northern Nigeria. Immediately after take-off it encountered a sudden wind reversal as it penetrated the 'cell' of a thunderstorm developing invisibly in its path. If the airport had been provided with the appropriate radar system this accident might have been prevented. It was not. Nor were the Argonauts fitted with weather radar. The latter had become available but the expense of installation in aircraft so close to the end of their economic life was not considered to be justified.

CHAPTER 12

GOODBYE TO PROPELLERS

There was undoubtedly a pecking order among the BOAC aircraft fleets in the 1950s. For a couple of years the de Havilland Comet 1 displaced the Boeing Stratocruiser in the prestige stakes even though it did not possess the range to cross the Atlantic. Then the tragic series of accidents interrupted the jet age. The Stratocruiser, the slow but stately piston-engined machine which could reach New York in seventeen hours again resumed pride of place unchallenged by any other airliner on the North Atlantic. The Lockheed Constellations flew east out of London to Australia and south to the Union of South Africa. Slowest and noisiest, the Argonauts served East and West Africa, India, Hong Kong and Japan. Shortage of aircraft had resulted in the total abandonment of services to South America in 1954.

The Vickers Viscount, operated by British European Airways, had been the first turbo-prop airliner in service in the world and had been ordered by many foreign companies. For BOAC Sir Miles Thomas ordered the Bristol Britannia, faster than any airliner currently in service. If it had been delivered on time it would have been another indication that the British could build a fine civil aircraft. As ill luck would have it the prototype made an emergency landing on the shores of the Severn estuary and its introduction was delayed for several years. The engine chosen for the Britannia was the Proteus which had originally been designed to power the Bristol Brabazon and the Saunders Roe Princess. These huge machines never attracted any orders at all and were scrapped. Before the Britannias were delivered to BOAC test flights revealed that the engines were liable to lose all power after entering cloud. The manufacturers sent a Britannia with a team of experts to Entebbe in Uganda. I recall the worried faces of these men as they pored over their blueprints, their aircraft returning from one test flight after another reporting "flame-outs."

When Sir Miles Thomas handed over to Sir Gerard d'Erlanger in 1956 the Britannia had still not entered service and the Boeing Aircraft Company were pushing ahead with plans to deliver to Pan American Airways their pure jet 707 airliner in competition with the strengthened and lengthened de Havilland Comet 4. All who came into contact with Sir Miles were impressed by his dedication and enthusiasm. He found the time to talk to the mechanics in the hangars and would travel in the tourist class and ask the passengers if BOAC were living up to their claim to "take good care of you."

I became aware of his shrewd style of management when I flew an Argonaut on the last stage of a flight from London to Lagos and was given a letter by the incoming Captain. This asked me to ensure that a particular passenger was well cared for and had no complaints. As it happened he was asleep when I went to see him. Later I learnt why the Chairman was concerned. The passenger was a shipbroker. Some weeks earlier he had arranged to meet a client on the latter's arrival at London Airport to negotiate a sale. He had telephoned BOAC to enquire about the arrival time of their aircraft and was given erroneous information, failed to meet his client and lost the sale. In his anger he had demanded compensation from BOAC and threatened legal action in pursuit of his claim.

Sir Miles naturally wished to settle out of court and instructed a young member of his staff to invite the shipbroker out to dinner to discuss terms. Several meetings took place and the young man attended the last of these with a cheque in his wallet which the Chairman had instructed him to offer at an opportune moment. Before he could do so the shipbroker made an offer of his own.

"I am surprised that your Chairman has entrusted this matter to someone as young as yourself," he began. "Presumably it would do your career no harm if I did not push you too hard."

The other agreed that this was true.

"Very well. You tell Sir Miles that I want a couple of first class return tickets to Lagos and I will settle for that."

The Chairman's protégé did not have to produce the cheque and BOAC were saved a not inconsiderable sum.

In 1957 the Britannia finally entered service and the strengthened Comet 4 was soon to compete against Pan American Airways on the North Atlantic where the latter were hoping to introduce the Boeing 707 ahead of the challenge by BOAC The pecking order was changing and we on the Argonaut fleet were amused to observe the disgust on the faces of the Stratocruiser crews when they found themselves swallowing anti-malarial tablets in West Africa instead of highballs in Manhattan.

I applied for a posting to Britannias in the expectation that these rather than the Comets would resume the abandoned routes to South America. I was wrong but it was soon clear that an improved Britannia would be put into service on the North Atlantic and Caribbean routes so I elected to remain on the Argonaut until this version was delivered. By the time I began the conversion course I had spent the last eight years on this fleet.

We had been obliged to cope with innumerable problems arising from complex political causes. In Egypt Colonel Nasser ousted King Farouk who sailed off to Rome in his royal yacht. In the company of his mistress he was often to be seen in one or other of the bars favoured by BOAC crews. When Nasser put pressure on the Eden government to remove the troops from the Canal Zone local sentiment was so aroused that we had to travel to and from the airport in unmarked buses with an armed guard. In 1956, in the aftermath of the British military action at Suez, services to Egypt were ended and we were not permitted to fly over the country. Flights to East and South Africa had to be routed through Libya which was still ruled by the friendly King Idris.

In Kenya the Mau Mau rebellion was of much more concern to the European residents than to transient flight crews but the sight of both male and female civilians carrying firearms as they went about their business in Nairobi made it clear how dangerous life there had become.

In Iran a precursor to the Ayotallah called Dr Mosadeq drove the Shah into exile and nationalised the Anglo Iranian oil company without compensation. We continued to fly to Teheran via Tel Aviv but had to carry sufficient fuel for the return journey to avoid uplifting oil whose ownership was in dispute

A different situation existed in Nigeria where the impending transition to independence was being greeted with great enthusiasm. Argonaut crews usually resided in the Ikeja Arms Hotel run by Mr Joe Harold who described himself as an "old coaster." He was on excellent terms with both the Nigerians and the white businessmen who worked there. No Europeans could own land in the country as was the case in Kenya. Meanwhile ministerial posts had already been awarded by the Governor to prominent local politicians.

Joe was a shrewd old fellow who correctly foresaw that independence would lead to tribal rivalries and a lot of trouble. He told me that the representative for Leyland buses had arrived from England to present a tender for a new contract to the recently appointed minister concerned with transport. Joe had asked Leyland's man if he expected to win the order.

"Of course!" he was told. "Our buses have served Nigeria very well indeed for decades."

"The contract went to Mercedes," Joe concluded, "and the ships which brought them here also put ashore some gleaming black saloon cars for the new minister and his closest associates."

One of my last trips to East Africa on the Argonaut was to Dar-es-Salaam where we spent a torrid night under mosquito nets in an old hotel cooled only by fans. Before turning in I was sipping a final iced drink and listening to our stewardess bemoaning the imminent approach of her thirtieth birthday.

"Still no husband in sight and I shall be asked to leave BOAC soon," she confided. "I am going to enquire about vacancies in East African Airways. I know Kenya has a poor record when it comes to settled matrimonial bliss but what is a girl to do?"

When I saw her the following morning her morale was at rock bottom. During the night an intruder to her room had been interested only in stealing the purse which she had put under her pillow.

"That's the final insult," she asserted, vehemently fighting back the tears. "I was absolutely starkers because of the heat but my money was all the thief wanted."

On that same trip there was a farewell ceremony on the airport in honour of the departing Governor. The skies on the route to Nairobi were cloudless and I made a diversion towards Mount Kilimanjaro so that all on board could have a close view. The passengers were informed and I did a complete circuit so that my co-pilot and anyone else who cared to do so could take photographs. The Governor told me that he had been flown around the mountain before "but never quite so close." When I was given a copy of the photograph taken by my co-pilot I was so impressed that I suggested that the editor of *"BOAC News"* might wish to publish it.

"That was *my* first thought," he answered, "but then I visualised some awkward management type reflecting that Kilimanjaro was not on the approved airway and remembering that years ago an airliner had flown into it when it was hidden in cloud. I think we had better leave well alone."

There had been some changes to the crew complement of the Argonaut. Improved radio equipnent and communications facilities eliminated the need to transmit and receive in morse code. The radio officers became redundant while the pilots took over the continuous 'watch,' sitting with their earphones bringing a constant flow of messages from controllers on the ground to all the aircraft within their responsibility. The non-pilot navigators were also replaced as there emerged from the Airways school at Hamble a flow of young pilots whom we called 'Pin' boys, 'pilots-initially-navigators.'

When I made my last flight on an Argonaut in May 1958 I had flown 5,500 hours on this type which represented half my total hours flying. A few days later I started the course of ground studies for the Britannia. This continued for five or six weeks and was exceptionally thorough. The aircraft had about three tons of electrical equipment on board and the electrics instructor with his numerous charts bombarded us with information which I was convinced was far more than for all practical purposes a pilot needed to know.

Having spent a bewildering morning looking at a chart which rather resembled a map of London's Underground system I asked the other ten pilots on our course if they were as bemused as I was and if so whether they would support me if I asked the instructor to confine the instruction to the basic facts, allowing us to learn how to interpret faults when they were revealed and to use the equipment to correct them. Most enthusiastically agreed that I should do this. After the lunch break I made a polite little speech; the instructor's face turned puce and he asked if that was the opinion of us all. No voice rose in my support.

"I'm sorry," I told him. "I have managed to get the drift of the engines instructor and the hydraulics instructor but I'm sure you are going too deep. I cannot even begin to imagine what a solenoid or a switched feeder look like and we are treated to many other such words which to you are as commonplace as apples and pears. Not to me, I'm afraid."

Everyone laughed and the rest of the instruction became more intelligible. At any rate we all passed the examination which was the same as that which our licensed flight engineers had to take.

By this time a new examination called Performance 'A' had been devised. For many years each type of airliner had been assigned a maximum all-up weight for take-off and another for landing. Airline pilots flying on international routes were aware that take-off on a cold day from an airport runway near sea level presented no problems whereas it was a very different matter taking off under an African sun from an airfield 3,000 feet or more above sea level. An engine failure at a critical moment as the end of the runway loomed was not a happy thought to contemplate.

When the licensing authorities decided to tackle this matter they went a lot further than consideration of a power loss on take-off. They took into account the possibility of high ground on the aircraft's track in addition to the runway length, its gradient, the ambient temperature, the wind component and of course the height of the airfield above mean sea level. The landing phase was also evaluated. It was necessary to know at what weight an airliner crossing the threshold of the runway at fifty feet could be halted within safe distance in varying conditions of temperature and so forth.

Certainly this study was overdue. It was eminently important for a pilot to be sure that he was operating within the limits of safety. In due course the technical manuals for each aircraft type were provided with graphs so that answers could speedily be found. But as is the way with examiners pilots were required to learn how to construct the graphs. Rather like the letters Q.E.D. at the end of a problem of geometry we were instructed to write "W.A.T. (weight and temperature) not limiting" when we had written down our answers to the various problems set. At the conclusion of one of these I forgot to do so and had to take the examination a second time.

The next hurdle was instruction on the flight simulator, a recreation of the aircraft flight deck complete with the hum of engines and the pitches and rolls associated with turbulence. When we became accustomed to the normal handling responses the instructor introduced every imaginable sort of equipment failure from a defective bulb in the undercarriage warning system to the loss of half the hydraulic supply. After a couple of weeks of that it was like a holiday to drive south to Branksome Towers Hotel near Poole and report each day at Hurn to fly the Britannia itself.

I made my first flight to North America at about the same time as Comets and Boeing 707s entered into competition across the Atlantic. Neither of these two airliners could fly to New York non-stop against the prevailing wind but the Boeing company could be counted upon to develop improved models which would do so. Sadly de Havilland did not have the financial resources nor a long enough order book to develop the Comet further. BOAC had no option but to place orders for a later version of the 707 to be powered by Rolls Royce engines.

The Britannia could usually make the direct flight to New York in about ten and a half hours but prudence obliged us to refuel en route when the winds were exceptionally adverse. Shannon airport with its large duty-free shopping area was popular with passengers, Gander less so. The fleet network of routes embraced Detroit and Chicago in addition to Boston and New York in the United States. In Canada only Montreal and Toronto were served. We also flew from London to Bermuda and thereafter to Barbados, Trinidad, Caracas and Bogota. Jamaica and Nassau were visited daily from New York.

My crew were enjoying a break in the delightful and unsophisticated Morgan's Harbour Inn in Jamaica at the time when scenes for the first James Bond film *"Dr No"* were being shot there. Rather to our surprise the film unit demanded an unusual amount of cooperation from the hotel guests who were asked not to make any noise in the poolside bar and not to move or use certain chairs and tables facing the ocean when shooting was in progress. The film unit's public relations officer had a difficult time with guests who objected to restrictions being placed upon them.

"What sort of crummy grade 'B' movie is this anyway?" an indignant voice asked.

"I assure you, it is a major production."

"Who is this Sean Connery guy anyway? We have never heard of him stateside."

"Drinks at the bar will be on us when each day's shooting is completed."

That had its effect, as did the appearance of the leading lady, Ursula Andress. Elderly male guests suddenly developed an interest in film-making. The younger couples happily accepted the invitation to appear as extras in the dance floor scenes. Some wilted as the director continually ordered retakes.

As more Boeing 707s appeared in the skies over the North Atlantic Britannias became increasingly relegated to charter work. The comfortable first class cabin with leg rests and ample space disappeared and extra rows of seats were installed to conform to the economy class configuration. The departure time from Heathrow was invariably 10 pm or later. By fitting a couple of bunks for the use of the flight crew the limits on flying duty could legally be extended and my log book records a trip in October 1960 when I flew from London to New York with stops at Manchester, Prestwick, Montreal and Boston. The sector from Prestwick to Montreal took nine hours forty five minutes. Each of the others lasted just under an hour. After such trips we were allowed two consecutive nights rest in our hotel to recover.

A long haul but less of a marathon was the ten-and-a-half-hour flight from Heathrow to Bermuda. As this airport had only one runway which we could use and there was no other airport within our range it was necessary to calculate the point along the track from England after which it would no longer be possible to divert to Gander. Fortunately I was never obliged to do so. Passengers eager to escape the northern winter were unprepared for an unscheduled stop in Newfoundland and did not appreciate the vista of snow.

Our local manager on the island had worked for Imperial Airways and thought it fitting to put the Captains in the Mid Ocean Club which had its own golf course, tennis court and private beach. The remainder of the crew were lodged in an hotel in Hamilton, the capital.

There were seldom less than four Captains passing through Bermuda at any one time and as most were keen on golf or other sporting activities they were happy to stay at the Mid Ocean. During the winter we were almost the only guests and the staff and waiters were quite pleased to have visitors to talk to. When the weather improved and every room was occupied by big spenders and generous tippers we had great difficulty catching the eye of a waiter or barman. We were also unique in being the only golfers who did not engage caddies. Their fees were so high that some of us wondered whether we were in the right job.

The club professional, Archie Compston, was getting on in years but could still make par on the course. He was particular about his partners but would play with Bermuda's Governor. Following them one day we waited until they had passed out of sight over rising ground on the fairway and when we thought we had allowed them sufficient time to be out of our range we drove off. Four respectable drives soared away and as we strode after them we met the Governor heading back towards us to complain that he had almost become a casualty from our cannonade.

Archie Compston was a martinet towards his pupils and charged them extortionate rates on the principle that he would not be considered in the top class if he failed to do so. I watched him trying to teach an elderly woman whose shots on the practice range went in every direction but never very far.

"Madam, can you cook?" he asked at last.

"Oh! yes. Ask my husband: I am an excellent cook."

"I would recommend you to stick to that then."

At that time Bermuda had strict rules in regard to dress. Shorts could be worn, even after dark, when jackets and ties were *de rigeur*. But they had to be knee-length. Notices in hotel bedrooms informed lady guests that only one piece swimsuits were permitted in the pool area. A stewardess was seen to be swimming in a bikini and an agitated pool attendant hurried over to her male companions and told them about the rule.

"Please tell her that she must confine herself to a one-piece costume."

"Willingly," he was told. "Which piece would you prefer her to remove?"

For me one of the delightful aspects of airline life was to fly away from a country wrapped in cold mists or under feet of snow and be able to spend the evening several thousand miles away sipping a gin and tonic under a tropical sky. One could never count upon it however. On a day in Bermuda when the wind had rattled the window panes and the rain had been unceasing I telephoned to my crew at their hotel and suggested a visit to the cinema in St George's parish. Only the stewardess turned up and I must confess at once that the others missed a really dreadful film. When the two of us walked out some time before the end of the programme the skies had cleared. Under a mass of stars we walked along to the jetty on the sea front. To my astonishment a Royal Navy submarine was berthed there, a solitary sentry standing guard by the gangplank.

"Shall I ask if we can go aboard?" I asked rashly.

"Yes; lovely if we can."

The sentry saw us approach and glowered. "Another couple of bleeding tourists," his expression seemed to say. "A bit late isn't it, squire?" he replied to my request.

"Perhaps you would present my compliments to the officer of the watch and enquire whether we may be permitted to board?" I swallowed hard at my own effrontery and mentioned my name and rank.

The sentry disappeared and returned almost at once with a lieutenant whose scowl suggested that he was not going to put up with any silly jokers. Then he took one look at my companion whom it must be said was a magnificent advertisement for BOAC and welcomed us aboard. A sub-lieutenant writing a letter in a wardroom of diminutive proportions was promptly ordered to show me over the ship. Regretfully I followed him as his superior produced a bottle of Gordon's gin and two glasses, eager to devote himself wholeheartedly to his unexpected female guest. The sub-lieutenant was determined that I should be shown every nook and cranny of the submarine. There were an awful lot of these and the periscope was raised and lowered a number of times before we returned to the wardroom. Judging by the level of the gin in the bottle the lieutenant and the stewardess had been getting on splendidly in my absence. Before we took our leave I was presented with the entire ship's mail to post on my arrival in England.

"That would be awfully decent of you," the lieutenant said, "seeing that you will be back home in twenty four hours or so."

"Did you enjoy the evening?" I asked the stewardess a little later.

"Absolutely marvellous. I have always had something of a yen for mariners."

The Britannia fleet manager was a most amiable Scotsman called Captain Anderson. From time to time he arranged a meeting which any Captains who were at home were welcome to attend. Frequently there was a guest speaker and any subject could be discussed. One such meeting was addressed by BOAC's security chief. Heathrow has been called "Thief Row" for the considerable amount of crime committed there. The payroll for BOAC's weekly-paid staff was once hijacked by a gang in smart business suits who drove off the airport through a gate on the perimeter which was normally padlocked and intended only for the use of the emergency services. The considerable sum stolen was then used to organise the Great Train Robbery not long afterwards.

We were told that it was fairly simple for dishonest loaders working in the confined space of many airliner cargo holds to slash mail sacks or suitcases and steal the contents. It was made easier for the thieves by the Post Office putting registered mail in distinctive sacks. Moreover as all consignments of gold or diamonds were insured the despatchers took a relaxed attitude to the losses that occurred. We listened in varying states of surprise and shock to these revelations and then it was our turn to ask questions.

"Surely BOAC, or the airport authority if they hire the loaders, can check to find out if an applicant has a criminal record before he is engaged?"

The security chief shook his head. "The police are specifically barred from disclosing any such information. Legally an individual's civil rights would be infringed by such a disclosure. I am not saying that it is impossible to obtain information from a serving police officer with whom one has worked in the past. But he is committing a serious offence if he tells us what we need to know."

I was in England when a public enquiry opened to consider the circumstances surrounding the crash at Southall, near Heathrow, of a Vickers Viking freighter. I attended the hearing because I had known the Captain and had made my first trip to Teheran with him on an Argonaut. He had lost his job with BOAC and joined a charter company operated by a colourful character called Captain Kozubski. The latter had earlier made the headlines when the Douglas Skymaster which he was piloting strayed over Albania en route to a Mediterranean destination and had been forced to land by MiG fighters.

Briefly the Viking freighter, whose load was two Proteus engines for El Al in Tel Aviv, left Heathrow after an extended mechanical delay. On its way to Israel an engine failed and the pilot was unable to keep the Viking from crashing as he tried to fly back to Heathrow. The enquiry revealed that the Viking's weight was above the permitted level when it took off, the flight time limitations of the crew had been exceeded and two men who had worked on the engine which had failed were not licensed engineers.

I watched and listened with interest and particularly noticed that counsel representing the charter company was far better briefed than those representing the dead pilots and other parties.

The latter positively floundered in their ignorance of the technical points being raised. Someone sitting very close to the company's counsel could be seen prompting him at every stage. Suddenly this man turned his head in my direction. Our eyes met and he gave a broad grin. He was dear old George of the perfume and currency rackets.

A little time passed and I read his name in connection with a case which opened at the Old Bailey. A bag containing prohibited drugs had been found on an airliner on its arrival at Heathrow. Although George was not a member of the crew which flew the aircraft on its final stage to England he had been the pilot earlier and he stood in the dock accused of responsibility for the carriage of the bag. George was acquitted very speedily, the judge ruling that the court had no jurisdiction to proceed to judgment as the actual offence was probably committed abroad. Counsel for George declared that if the case had been continued his client had irrefutable proof of his innocence.

My last flight in a Britannia was a charter to Buenos Aires carrying members of the Band of the Irish Guards and Royal Scots Pipers. On their arrival they marched to the statue commemorating Simon Bolivar who had made a great contribution to the independence movements of Argentina and other South American countries in the early nineteenth century. My crew then flew north to Recife on the coast of Brazil where another crew took over and we were accommodated in a pleasant hotel facing a splendid white beach, ideal for swimming. It was November 1963 and two days spent in that relaxing atmosphere prior to my return to an English winter and a conversion course to the Vickers VC-10 was most appealing.

After dinner in the hotel we went for a walk along the front and visited an open-air market doing a good trade under trees with fairy lights. Then being attracted towards a building from which the sound of dance music could be heard we strolled over to investigate. It was not clear whether it was a private party but shouted invitations to join them from cheerful revellers on the verandah were accepted by a few members of my crew. The following morning we were all lying on the beach when some dusky Brazilian maidens appeared and flung themselves down alongside us.

"'Ullo Johnnie! 'Ullo Reggie!" they cried, and several faces were observed to turn bright red as the rest of us asked to be introduced.

I had enjoyed the five years which I spent on Britannias but it was not a period which had been financially rewarding for BOAC. Sir Gerard d'Erlanger who had been Chairman until 1960 complained that having placed orders worth £150 million for

British aircraft huge extra charges had been incurred through necessary modifications before they could be put into passenger service. Such development costs did not arise when BOAC bought American aircraft. He believed that the extra costs should be charged against the funds which the Government made available for research and development.

In this respect the Britannia had proved very expensive to operate. But by the time they were sold the engines were performing so well that the time between overhaul had been greatly extended by the licensing authorities. Their new owners incurred far lower operating costs than BOAC.

When Sir Matthew Slattery succeeded d'Erlanger seventy per cent of BOAC's airliners were of British manufacture and about ninety per cent of the engines. But just as the corporation had greatly increased the number of seats on offer on its worldwide network the travel industry entered a period of nil growth. Comets and Britannias due to be phased out were rapidly losing their book value. Seventeen individual companies ranging from wholly owned subsidiaries to foreign airlines in which BOAC had minority holdings were also costing the corporation dearly. An investigation into these financial problems was ordered by the Minister for Aviation. It was never published but Sir Matthew Slattery resigned at the end of 1963, along with the Managing Director Sir Basil Smallpeice.

CHAPTER 13

THE JET AGE

Although some of my contemporaries had transferred to the Boeing 707 fleet I was content to remain on the Britannia until pilots were invited to apply for conversion to the Vickers VC-10. Unlike the misfortunes which beset the development of the Tudor, Hermes and Britannia the initial flight trials of the prototype VC-10 proceeded satisfactorily. However the fuel consumption was found to be higher than expected. In normal circumstances the aircraft was not going to be able to carry the same number of passengers as great a distance as one of BOAC's Rolls-powered 707s. The reason for this was the drag induced by the massive T-tail. On the other hand the VC-10 had been designed to meet the airline's requirement that it should lift a full load out of those airports situated at high altitudes and prone to high temperatures. Nairobi and Johannesburg were two such places. Here the VC-10 outperformed both the Boeing 707 and the Douglas DC-8.

From a practical viewpoint the pilots' conversion course was sensibly directed at facts which we needed to know. Our engines instructor was a droll character with a dry sense of humour.

Frequently having to interrupt his discourse as one flight after another roared out of Heathrow he was heard to mutter that "this job would be all right if it wasn't for the aeroplanes." Caught out in some minor error of fact he asked sadly: "Well, what can you expect from someone paid £1,500 a year and all the chalk he can eat?"

When I had been engaged on the ground studies for the Britannia six years earlier I had noticed that the younger pilots absorbed the information being imparted to them more readily than their seniors. Most of the Captains were men who had joined the RAF during the war. Aptitude in the mechanical sciences had not then been demanded. After the early 1950s the numbers of former service pilots dwindled to a trickle and the Airways Corporations opened their own flying school at Hamble. The pupils whom they recruited were those with qualifications in engineering sciences, mathematics and physics. This background served them well on the VC-10 course. It only required some bright young fellow to develop an argument with the remark "then the figure varies with the cosine of . . ." to leave me feeling like some ancient mariner from the era of sailing vessels confronted with the complexities of the newfangled steamships.

Success in the written examination was followed by hours spent on the VC-10 simulator before we were sent to Shannon airport for a week of flying. Just as Gander had ceased to draw in airliners obliged to replenish their fuel tanks, so too had this airfield in southern Ireland, and flying training could be practised uninterrupted by the arrival and departure of commercial flights. All of us were enthusiastic about the handling qualities of the VC-10 and eager to fly it in passenger service.

But now a shadow blighted the prospects. The new Chairman, Sir Giles Guthrie, was only too aware of the financial problems of BOAC, problems which had terminated the chairmanship of his predecessor. If Comets and Britannias had to be sold at knockdown prices while huge sums had to be raised to pay for a fleet of modern jets it made more sense to him if the new aircraft were ordered from Boeing with the advantage of compatability in spare parts to be shared with the airliners already in service. He obtained an assurance from the Minister for Aviation, Julian Amery, that the Government would compensate BOAC if the latter was compelled to accept VC-10s against the airline's commercial judgment.

Although a compromise was worked out and some orders for the Vickers VC-10s were cancelled the wrangle did nothing to enhance Vickers' hopes of foreign sales. The small number of VC-10s ordered from all sources was a bitter blow to the manufacturers who abandoned their plans to develop the type beyond the Super VC-10 which went into service on the longer sectors. Sir Giles Guthrie was more fortunate than his predecessors. The slump in the air travel industry ended soon after he assumed office and 1964 witnessed very high load factors on many of BOAC's routes. An exception was the east coast route to South America which was abandoned once again when the Comets were withdrawn.

After completing the VC-10 course I moved to a village in Oxfordshire close to the aerodrome at Brize Norton where I had been stationed for a short while in 1942. When I took possession of a delightful Cotswold stone cottage a bomber squadron of the United States Air Force were moving out from their base but within a year the RAF returned there and in addition to a squadron of Short Belfasts a VC-10 squadron was formed. The Station Commander, Group Captain Bob Wilson, had been a pilot for as long as I had. His wife Margaret had once been secretary to Mr Whitney Straight, Vice Chairman of BOAC. Little as was my experience of VC-10s that of his own pilots was less and Bob used to call in at my home to discuss any problems which were being encountered.

I was invited to become an honorary member of the Officers' Mess and recall one festive occasion when much wine had flowed. I was challenged to justify the higher salary paid to BOAC pilots for flying the same aircraft as the pilots in the squadron. The officer who confronted me was almost offensively persistent in demanding an answer.

"Don't blame me for whatever you are paid," I replied, and rather rashly added: "It is not my fault if the RAF are the Pakistanis of the aviation scene."

There was a shocked silence before someone said: "He is right, you know. We are the worst paid air force in NATO."

I moved swiftly to avoid further discussion of this topic.

My choice of residence reflected the airline's requirement that crews on stand-by should be available at the airport at ninety minutes notice. To comply with this I kept a packed suitcase in my car and an adequate supply of petrol in the tank. When I got up in the morning I put on my uniform. I was called out several times and always made the journey within the stipulated period but one such occasion remains in my memory. Soon after midnight I was woken by the Duty Operations Officer. He was most apologetic.

"I suppose you need me for one of the earlier flights tomorrow?" I said.

"No. We need you now," he replied.

"But jet aircraft aren't allowed to take off after 11 pm," I reminded him.

"We have been given a dispensation," he told me. "Our last flight to New York has had to turn back and we have a VC-10 available to transfer the passengers and get them away again. The only restriction relates to weight. To avoid excessive noise you will have to take off light and refuel either at Prestwick or Shannon."

There was little traffic on the roads to Heathrow and I took over a thoroughly disgruntled group of passengers. We reached New York after a refuelling stop at Prestwick and at about four o'clock in the morning local time my crew started the journey into Manhattan and our hotel. To my amazement the road was choked with traffic.

"What on earth . . . ?" I began to enquire of our driver. "The bus companies and subway system are both on strike, Mac," I was told. "These poor guys are using their own cars in order not to lose a day's pay."

By the 1960s the professional standard of all BOAC pilots was scrutinised each year on three occasions. A training Captain would accompany a crew on at least one sector of a regular passenger service. A competency check would be conducted on the simulator which offered an opportunity to an instructor to introduce a number of different malfunctions of equipment and emergency situations. These included an engine failure on take-off or a fire warning in order to refresh memories of the correct drills and ensure that they were satisfactorily performed. The simulator was also a practical device for the renewal of the instrument ratings of pilots besides saving fuel and other costs. One would be expected to complete a 'flight' using the instrument landing system (ILS) when the visibility and cloud base were at the minimum legal limits.

This last exercise was not supposed to be complicated by the introduction of malfunctions or emergencies but such is the diversity of human nature that I remember one training Captain who recorded as "failed" a series of pilots by setting them a trap over the altimeter setting. The reading of this instrument varies with the barometric pressure. Heathrow is not many feet above sea level. The pilots have to reduce engine power about ninety seconds or so after beginning their take-off run in order not to exceed the tolerated decibel limit as the aircraft passes over a listening post. During this time the undercarriage and flaps have to be retracted and the instructions of Departure Control answered. Reaching 4,000 feet the altimeter has to be reset to 1013 millibars to ensure safe separation of all airliners at and above that level. By giving the pilots an extremely low altimeter setting prior to take-off and a clearance to a flight level of only 4,500 feet the training Captain expected that by the time the pilots had reset the altimeter the aircraft would have passed a few hundred feet above the level to which the 'controller' had cleared them.

So many of us fell into this trap and had to return on another day for a further check that the roster clerks had difficulty in assigning pilots to flights. When the Fleet Manager was told the reason he complained to the examining Captain that he did not wish instrument rating renewals to be conducted in such a manner.

He received a cool response.

"How I perform my duty is entirely a matter between myself and the Minister," was the reply that he received.

It would be hard to find a long-haul pilot who preferred to hand-fly his aircraft rather than make use of the automatic pilot but the introduction of a flight system which could be coupled to the autopilot to effect a 'hands off' landing in conditions of minimal visibility, whilst a wholly desirable development from the airline operator's point of view, was viewed with considerable suspicion by some Captains. We had been asked to use the equipment in good weather conditions to build confidence in its performance but inevitably there had been occasions in the development phase when the autopilot behaved erratically or simply disconnected, leaving the pilot to take over and complete the approach and landing manually. It was not unknown to hear a Captain declare that so long as he was responsible for his aircraft he would not rely on the equipment to perform the landing. Only in the late 1970s was it so improved that the Boeing 747s and Trident 3s of British Airways were certified for landings in very poor conditions of cloud base and visibility.

It had been a longstanding grievance of the families of crew members that it was difficult to obtain information from the airline when we did not reappear on the days when we were expected to return home. Serious domestic problems arose when a wife was incorrectly assured that her husband had indeed returned from his flights when in fact his crew had been rescheduled for some reason. Eventually a genuine effort was made to keep families properly informed and thereafter far fewer complaints were heard. Inevitably with so great a number of individuals and commonplace names like Smith or Robinson abounding errors still occurred. The wife of a friend of mine was woken by the telephone to be notified that her husband's aircraft would be landing in a couple of hours.

"Really?" she replied acidly. "In that case I had better take a closer look at the man who is snoring in my bed upstairs."

"I am so sorry," the flustered clerk answered. "Hang on and I will recheck. I may have made a miscalculation."

Indeed she had and was subsequently dubbed Miss Calculation. I remember her as a gentle soul who prepared the pilots' roster and would always try and help if one wished to be in England for some special family occasion.

Another kindly lady called Mrs Mack prepared the roster of stewardesses. One of the latter mentioned to me that having been taken ill very soon after returning from abroad she visited her own doctor. He asked her where she had been immediately prior to her return. Then he enquired whether she drank much milk. Indeed she did. She liked milk and drank quite a lot of it.

He told her that she had probably contracted brucellosis and gave her a letter to pass on to her employer when she felt fit to continue flying. The stewardess took the note to Mrs Mack who expressed pleasure that the girl had swiftly recovered and then perused the spidery handwriting of the doctor. The latter had defined the complaint by its alternative name of abortus fever and it took the two women a few moments to decipher the words.

"Now don't worry, dear," Mrs Mack said. "My lips are sealed. This will remain a secret between the two of us."

A third roster clerk was known as Dracula's daughter. Inured to the fervent protestations of crew members who found their social life ruined by a sudden command to report in for a training session or other unwelcome duty and knowing that she was in effect the messenger of management she was unyielding. But she too could miscalculate, as on the occasion when she demanded to know why a certain Captain had not reported in for a simulator exercise.

"It could be because he is somewhere in the Pacific Ocean," his wife replied. "So it would probably be easier for you to contact him than for me."

Dracula's daughter had a warped sense of humour. It amused her to assign to Captain "Corpus" Deadman a co-pilot called Mort. The former once signalled to his destination airport a request for an ambulance to be available on his aircraft's arrival. During the flight a passenger had been taken ill. After the landing and when the engines had been switched off "Corpus" could see no sign of an ambulance, neither did a doctor or nurse accompany the traffic officer on board.

"Didn't you receive my message?" he asked impatiently.

"Certainly, Captain," he was told. "About the dead man. The morgue would not send out an ambulance but don't worry. Their van is here."

In the early post-war years BOAC regulations had denied access by passengers to the flight deck unless a pass had been issued by the Chairman in writing. Only once was one of these presented to me. But Captains receive an inordinate number of requests to visit the cockpit and do not wish to be distracted from their work except at quiet times on the longer flights.

Consequently the decision was subsequently left to their own discretion and short visits by passengers, one or two at a time, have proved to be a useful exercise in goodwill. One evening approaching Bombay a steward told me that a passenger was very concerned that he might miss his connection with a Czechoslovakian airliner as we were running thirty minutes behind schedule. I told the steward to bring the passenger forward. He was an American citizen of Japanese origin. I pointed out to him the Czech airliner cruising 2,000 feet above us on the same course and explained that Air Traffic Control would call in aircraft from the bottom of the stack so that we would be bound to land before the Czech. He was most grateful for this information and then told me the following fascinating story about himself.

In December 1941 when the Japanese suddenly attacked Pearl Harbour and brought about America's entry into the war he had been on a visit from his home in California to relatives in Tokyo.

"I was a US citizen with a US passport but in the circumstances of the time it was not something it would have been wise to proclaim. I was called up for military service and became a pilot. When the Emperor ordered the armed forces to lay down their arms I had just been posted to a Kamikaze suicide squadron."

"What do you do now?" I asked him.

"I am Professor of Oriental Studies at the University of Michigan."

On another flight I was told by the chief steward that a passenger who had worked as a stewardess for Pan American Airways had asked to visit the cockpit. We were busy at the time and I said that her visit would have to wait. Some time later I remembered her request but my co-pilot was unenthusiastic.

"I have had a look at her," he remarked, and added, damningly, "It must have been a long time since she flew."

The chief steward reappeared to renew her request and I reported to him my co-pilot's disparaging observation.

"Captain," he replied. "At your time of life and mine we cannot afford to pick and choose. I would describe her as a good little runner with quite some mileage remaining."

Since the days when the stewardess was the only female member of the crew the attraction of the job has undoubtedly diminished. Serving and collecting the trays of food of over four hundred passengers cannot be called glamorous and in a Boeing 747 she will be one of about fifteen cabin attendants most of whom will be other women.

It is a hard task for a major British airline to find enough young women with fluency in any language besides English, combined with experience in dealing with the public as a nurse, receptionist or social worker. I remember girls who had been schoolteachers, dancers, dentists and veterinary assistants. Another had been a concert pianist and always travelled with her silent keyboard to keep in practice. Some were much more caring of those in their charge than others.

One of the former was in a cinema in New York. Her enjoyment of the film was being spoilt by the repetitive hacking cough of a woman seated directly behind her. Remembering that she had a tube of throat pastilles in her handbag she fumbled within it, shook out two tablets, turned round and offered these to the coughing woman. They were gratefully accepted and appeared to have some effect. Before long there was an intermission and the lights came on. Glancing down at the tube which she still held in her hand the stewardess was dismayed to find that she had parted with two tablets from a phial with which she fertilised the pot plants at her flat. Greatly concerned she rushed out of the cinema and found a chemist's shop, handed over the phial and asked if any serious harm might befall the unwitting consumer.

The chemist was soon able to offer an opinion.

"Don't worry," he counselled. "Your patient will have a very uncomfortable day but suffer no lasting ill effects."

"What is in those tablets?"

"Concentrated sheeps' manure."

More adventurous was another girl who flew with my crew from London to Dubai in the Persian Gulf. We landed at about three o'clock in the morning and were due to be picked up from our hotel twenty three hours later. When we assembled in the lobby she did not appear and the hotel porter was sent with a pass key to unlock her bedroom. He reported that the bed had not been slept in nor was her bag packed in readiness for departure. Puzzled and rather concerned we set off for the airport in the crew bus. I had forgotten all about her by the time I had signed the load sheet, started the engines and was almost ready to taxy out. Then I saw a stewardess, suitcase in hand, rushing out to board the VC-10 as the steps were about to be pulled away. We departed for Bombay and it was only when we were on our way back to Dubai some hours later that I remembered the incident. After I had been told that the passengers' breakfast trays had all been recovered I had an opportunity to ask the chief steward what explanation the girl had offered for her behaviour.

"It's her first trip and she was confused by the time difference from London, Captain."

"First trip! Not in her own bedroom at two in the morning! Has she family friends in Dubai?"

"She didn't say so. I told her that it was a serious matter and that you would deal with her yourself, Captain."

"Thanks very much," I replied. "Are you suggesting that I put her across the navigation table and smack her bottom?"

He nodded approvingly. "That's no more and no less than she deserves."

"Maybe so, but I have kept my name out of *"The News of the World"* so far and I hope to retire with a full pension. Send her up to me."

She was a fairhaired girl of about twenty with an innocent expression. In answer to my enquiry she repeated her story about confusion over the local time. Yes, she was out with friends.

"Did you know the people who took you out?"

She was evasive and it turned out that in the short time we had spent in the hotel an Arab had approached her and invited her to meet his friends. In their own countries the wives and daughters of

Arabs never set foot in hotels used by Westerners. Having daughters of my own of her age I was dismayed at the risk she had taken but decided not to labour that point to excess. I told her that if she had missed the flight it would not have gone unreported and that she would probably have been sacked.

"You won't be reported this time," I said, "but in future show more prudence."

One seldom hears or reads that word 'prudence' nowadays. I wondered if she knew what it meant.

I remained on the VC-10 flight until I retired shortly before my fifty fifth birthday in 1977. It was a delightful aircraft to fly and its quiet cabin was appreciated by the passengers although the noise from the powerful Rolls Royce engines was singularly unwelcome to those who lived or worked anywhere near an airport's immediate flight path. The VC-10 flew over most of the airline's route network and continued to do so after the appearance of the Boeing 747, the first jumbo jet, because some time passed before airport taxiways and buildings could be altered and enlarged to cope with an aircraft of that size.

We flew holiday makers to Caribbean locations such as Saint Lucia and Trinidad in the Caribbean. In these parts hotels were generous with rum punches and entertainments such as limbo dancing.

"Now let's see some of our guests have a go," our hosts would appeal, and if there was hesitation would cry out: "surely we have members of a BOAC crew to give you a lead." The challenge was invariably accepted.

Mauritius and the Seychelles came 'on line.' The airport at the latter was built on crushed coral deposited on the coast. This ended the island's dependence upon sea transport and allowed the creation of a tourist industry which flourishes in the dry season. Mexico City, situated over 7,000 feet above sea level and Bogota in Colombia over 8,000 feet were two of the highest airports served by the VC-10. Lima, capital of Peru, was added to the west coast network, neglected since the withdrawal of the Constellations in 1949. This gave me an opportunity when I was on leave to reach Chile for a visit to my relations, flying on to Santiago in the aircraft of my former employer LAN-Chile.

In addition to the flight from London eastward to Australia and New Zealand we enjoyed flying westward to Australasia through San Francisco or Los Angeles via Honolulu and Fiji. This great network flown by the VC-10 enabled me to visit from time to time my older brother who was a professor at the University of the West Indies and a younger brother residing in Perth, Western Australia.

In striking contrast to such pleasant places were others in the throes of civil war or recovering from them. The crumbling runway at Dacca in Bangladesh reflected the ravages of the recent struggle for independence from Pakistan. No ordinary passengers visited a country whose homes and crops were regularly inundated by flood during the monsoon season. The small number who disembarked from our aircraft were medical personnel and aid workers and our cargo similarly reflected the most urgent requirements of an overpopulated and poverty-stricken nation. As the large bins of discarded meals were carried out of the galleys by the local catering staff soldiers policed a queue of hungry people who sought to pick through the rubbish and lashed out with their staves at starving dogs and queue-jumpers.

Entebbe on the northern shore of Lake Victoria in Uganda had been one of my favourite places. Although situated precisely on the Equator it lies nearly 4,000 feet above sea level. The hotel overlooked the golf course which lay alongside the lake and one could depend upon the eager young caddies always to find one's ball. On all sides the tropical flowers and shrubs were a delight to any garden lover. Then the evil Idi Amin seized power and tried to curry favour among the Ugandans by expelling thousands of resident Indians. In so doing he denuded the country of just about every craftsman. Watchmakers, locksmiths, electricians and mechanics could no longer be found. The new regime instituted a reign of terror which scared away all the tourists and then the businessmen. Amin's so-called security police sometimes entered our hotel and took away guests at the whim of their leader. Some BOAC crew members who were arrested one evening were released hours later through the intervention of the British High Commissioner who had been instantly informed by a witness to the event.

On their return to the hotel the manager personally served them a meal.

"You are luckier than most of those of my guests who have been seized," he told them. "I have rooms full of personal belongings of arrested men. I certainly do not dare to enquire about their fate."

Soon after that Entebbe ceased to be used for a crew change and later still flights to Uganda were terminated. By then its economy was in ruins.

Addis Ababa, capital of Ethiopia, was added to the route network at about the time the Emperor Haile Selassie was deposed and all the royal family imprisoned. The first military dictator soon fell victim to other jealous officers. The country has continued under a military clique mainly concerned with suppressing the attempts of their Eritrean subjects to secede even while famine has stalked the land.

VC-10 crews were lodged in the comfort of the almost empty Hilton Hotel. Walking about on my first visit I thought I had found a public park but on approaching the open gates I was suddenly confronted by soldiers with sub-machine guns who angrily gesticulated at me to go away. I discovered that it was one of the former royal palaces. There was a curfew in force and if we chose to dine at a restaurant in town it was essential to order a taxi to collect us in good time. We were about to set off one evening when an Englishman approached me and asked if he might join us. He was making his first visit to Addis Ababa and did not know his way about. We made him welcome and over dinner he explained that he was a salesman for an American aircraft manufacturer and that his territory was most of Africa. He went on to say that his customers were invariably governments whose minister for transport was the person with whom he had to deal.

At that time much unfavourable publicity had befallen the Lockheed company through its payments to highly-placed individuals whom they had believed were in a position to influence the boards of their national airlines. The scandal had touched Prince Bernhard of The Netherlands and members of the Japanese Government. I invited our companion's comments on the affair.

"That sort of thing has always gone on," he remarked. "My company quote the purchase price of their aircraft but know perfectly well that they won't get it. They tell me the lowest figure they will accept and it is my job to get the best deal that I can. It has been my experience that, when I have made a sale and the contracts have been drawn up ready for signing, the official with whom I have been dealing will hesitate as he picks up his pen. Then I will be asked to credit a numbered account at a bank in Europe with some figure, say fifty thousand dollars. That's his cut of course. I am quite accustomed to that ploy. I get out my calculator and provided the new total sum is within my company's guideline I agree."

I was reminded of the description given to me of the sale of buses to Nigeria. The words amoral and unethical came to mind but remained unspoken.

If the salesman shared my thoughts he did not express them. He shrugged his shoulders and sighed. "It's the only way business can be conducted on this benighted continent."

The crushing military defeat of Egypt, Syria and Jordan by Israel, followed by the latter's occupation of Sinai and the West Bank of the river Jordan was to lead in due course to the hijacking of airliners belonging to the Americans or the Western European states. This form of violence was chosen by Palestinian terrorist groups to express their anger and frustration and also their sense of outrage that no powerful nation was willing to exert pressure on Israel to respect the aspirations of their Arab subjects.

BOAC flights through the Middle East were flown by VC-10s and on three occasions my colleagues fell victim to Palestinian hijackers. I had known Captain Cyril Goulborn since he and I had been seconded to BOAC flying boats in 1942. After the war he returned to BOAC from an RAF Sunderland squadron. In September 1970 in the course of one day the terrorists hijacked his VC-10, a TWA 707 and a Swissair DC-8, forced their pilots to fly to a landing ground in Jordan and, having waited long enough for the television reporters to arrive on the scene, released the passengers and blew up all three aircraft. Cyril and his flight crew, along with a British Major serving with one of the Trucial sheikhdoms, were kept captive in a part of Amman controlled by the Palestinians. When King Hussein was no longer prepared to tolerate the existence of an armed force within his Kingdom independent of his own rule he sent his troops to quell them. Cyril's captors fled as the soldiers approached but the British Major recognised the new danger to his fellow captives as the sound of gunfire grew nearer.

"If we are not careful the army will chuck a grenade in this building before rushing in themselves," he warned them. "We need to wave something white from a window and shout like mad for help."

They did as he suggested and several weeks of captivity were ended.

When the VC-10 commanded by Jim Ritcher was rushed by hijackers during a transit stop at Dubai they made clear their determination by shooting and wounding a receptionist and loader near the aircraft. At the time some passengers had left the aircraft to make purchases in the duty-free shop while others had remained on board. Jim was still in the BOAC duty room checking his flight plan. When the hijackers realised that they did not hold the Captain captive they let it be known that they would begin to kill the passengers unless he came forward. The airport's security officer urged him to keep out of sight rather than allow the problem to be transferred to some other place outside the control of Dubai.

"I had no wish to have on my conscience the deaths of my passengers," Jim told me.

He walked out to the VC-10 and, mounting the steps, was confronted by a hijacker holding a pistol to the head of a young New Zealander.

"Thanks for coming on board, Skipper," the latter said.

The crew were ordered to fly to Tunis where prolonged negotiations lasted some days. The hijackers demanded the release of some of their supporters held in gaols in Cairo and the Hague. Before the Egyptian authorities had agreed to release and send their prisoners to Tunis the hijackers shot a passenger and threw his body out of the aircraft. A girl became hysterical and losing his temper for the first and only time Jim shouted at the apparent leader of the terrorists that he ought at least to allow women and children to leave the aircraft. To his surprise they were permitted to do so. Soon after they had been released he noticed that one young woman was still on board, seated next to her male companion.

"They looked like hippies," Jim told me. "The sort who go out to Tibet or Nepal and imagine that drugs will be easily available and a more satisfying mode of life can be found. I asked her why she hadn't left with the others. She just jerked her thumb at her companion who was shaking with fear. She told me that he was at the end of his tether and would need her support. I admired her for that."

The nightmare ended when the Dutch government sent their detainees to Tunis and the prisoners of the Egyptians arrived from Cairo and were shown to the hijackers. Jim said that the latter were very pale and bore the scars of manacles on their wrists.

A third VC-10 was hijacked over the Mediterranean when two armed men burst into the cockpit. The co-pilot "Bunny" Warren was at the controls, the Captain being on the point of visiting the passenger cabin. The former recalled the grenade which was held at his throat.

"Glancing down I could make out the words 'MADE IN CZECHOSLOVAKIA' embossed on the metal."

The hijackers took the Captain into the passenger compartment and ordered Warren to fly the aircraft to Amsterdam. After the landing all the passengers were allowed to escape using the emergency chutes while the hijackers set the VC-10 ablaze, hurling bottles of the passengers' duty free spirits into the conflagration. Warren was in no doubt that these terrorists were strongly under the influence of drugs.

After these events many of the airlines and airport authorities throughout the world spent huge sums on detection devices to deter hijackers and on the employment of extra security staff. Consequently terrorists have altered their tactics, sometimes using dupes to travel with baggage containing an explosive device. Governments scrutinise the *bona fides* of agencies purchasing weapons for export and cannot be too careful about the ultimate user. At about the time I retired a Boeing 707 freighter was detained at an airport in the United States prior to its departure. Upon examination the contents of the crates in the hull were found to be armaments. The name of the Captain had a familiar ring: George was at it again.

My retirement date was fast approaching when I made my last flight to Australia. In Brisbane I played golf and was startled by the reptiles resembling small alligators which swarmed over the course. Called goannas they are in fact very large lizards. In Perth I played again at my brother's club where numerous kangaroos bounded across the fairways and kookaburras noisily responded to good shots and bad. Then on to Sri Lanka where the Royal Colombo Golf Club still proudly displayed a large portrait of Queen Elizabeth II despite the country's departure from the British Commonwealth of Nations. Nor was there any diminution in standards. Ice-cold drinks greeted the hot and thirsty players after their round and clean towels and hot water showers awaited them in the changing rooms. The only concession to the country's new circumstances was the presence of water buffalo tethered on the fairways. The club committee had drafted rules to afford players relief from that inconvenience. The course had a large stretch of water and any ball that fell within it was swiftly pursued by a host of individuals aged between seven and seventy keen to earn a rupee for salvage.

England was in mid-winter on the eve of my last VC-10 flight. Fog was forecast and rather than rise at some unearthly hour to reach the airport in difficult driving conditions I checked into a hotel near Heathrow. Breakfasting there the following morning I received a telephone call from Operations that my flight had been cancelled.

This seemed to me such a depressing way to conclude my flying career that I drove the short distance to the Operations centre and suggested that I took out some other service whose Captain had not yet left home. This was arranged and I took off for Toronto with my suitcase containing the lightweight clothing more suitable for the southern summer weather which I had expected to enjoy.

The temperature in Toronto was below freezing point and I had no inclination to spend any time out of doors. When it was time to fly home it was snowing and at the airport we joined a queue of airliners being de-iced before taxying on to the runway for take-off.

On the way home the chief steward brought forward a large cake prepared by the catering staff at Toronto. On the icing I was very touched to read the words "Happy retirement Captain Jackson."

CONCLUSION

A CHANGE OF COURSE

Many months before my impending retirement I had begun to give considerable thought to the subject of a future occupation. I had enjoyed most of the thirty five years which I had spent flying and the prospect of a long series of winters uninterrupted by flights to warmer climes was horrendous to me. The enjoyment of my garden does not extend to the months after September nor am I so fanatical a golfer that I will endure sleet and icy winds to reduce my handicap. I had come to realise that flying had been my principal hobby, the more attractive for being paid for it. Moreover I had written a couple of books on the subject which were about to be published.

It was during a stopover in Abu Dhabi that an idea was put into my head. An attractive stewardess of my crew who possessed a university degree expressed herself pretty bluntly.

"No one should be allowed to vegetate like a cabbage. When my father was made redundant I persuaded him to study for an Open University degree. He has been plugging away for it for some years now."

Open University? Correspondence courses? I was not excited by the thought but it reminded me that in the distant past I had obtained the much easier qualifications then in force for entrance to Oxford University, only twenty miles from my present home. Would an application for a fifty-four-year-old candidate be accepted?

My elder brother, an Oxford graduate who had worked as a lecturer in various universities, encouraged me to pursue the idea.

"Nowadays," he told me, "universities welcome the presence of mature students. Their experience in business or industry exercises a responsible influence on the youthful majority."

The friendly lady in Oxford's University Applications office hesitantly agreed.

"Well, ye-e-es, but mature students are somewhat younger than you, about twenty seven or so. However there may be some college which would consider you. As you still have some months remaining in your present job why don't you take an 'A' level to show that you are really serious?"

I decided to act upon her suggestion. History has always interested me and I enquired at Witney's College of Further Education whether they offered an appropriate course. Indeed they did and I enrolled at once and joined a very small class which included the wife of a police constable and a Polish girl who was trying to improve her English. I missed many of the classes through absence on flights but I obtained a copy of the syllabus, some past examination papers and a list of books and got on with the reading in my spare time.

I was fortunate enough to find a Master of an Oxford College who was willing to consider my application. He invited me to dinner and introduced me to some of his eminent colleagues in order to canvas their opinion before deciding.

One of them said to me: "Will you be awfully disappointed if you finally end up with third class honours?"

Subsequently the Master informed me that if I obtained a respectable pass in the 'A' level examination he would accept me.

I obtained a 'B' grade and the Master appeared more hopeful that he had not made a fearful blunder.

"Now, what languages will you offer? At the end of your first term you will have to take three examinations and two of them require the ability to translate separate languages with accuracy. How is your Latin?"

I chose Spanish and French and still recall the amusement of my crew as I waded through Tocqueville's *"Ancien Regime"* with the aid of a dictionary while we sat by a swimming pool in Colombo.

In October 1977 I joined a procession of young men and woman entering the Sheldonian Theatre and emerged shortly afterwards as an undergraduate. I continued to live at home and drove into Oxford most weekdays, often dining in the college before returning to my village. I was interested in most of the material which I had to read and as I happen to enjoy writing it was no great chore to produce and read to my tutors the essays which they had set. Sometimes they complained that they were too long. It amused me that most of the tutors addressed me as Mr Jackson whereas the undergraduates found no difficulty whatsoever in treating me as one of themselves and calling me by my Christian name.

Undergraduates have the choice of a very large selection of lectures which they are free to attend if they wish. Few were so popular that my presence went unnoticed by the lecturer. At their conclusion I sometimes received a polite enquiry about my identity. Once incredulity was expressed.

One good lady stopped dead in her tracks as she handed out some notes and told me: "I cannot believe that anything I shall be saying can possibly be of any interest to you."

Everyone present craned their necks to identify an interloper who had perhaps drifted into the building to find somewhere warm to pass the time.

"I am an undergraduate reading history," I said and she gulped in astonishment and blushed.

More amusing was a remark made in the course of a lecture in British economic history.

"Have a word with your grandfathers," the undergraduates were advised. "They will tell you that in the 1930s they belonged to a Friendly Society."

The words struck a chord. Thinking hard I remembered that as a seventeen-year-old working for a few months in a merchant bank in the City of London I had belonged to a Friendly Society myself.

In June 1980 I put on the traditional white bow tie, dark suit and undergraduate gown and joined a stream of others similarly attired to attend "schools" and to write nine examination papers over a period of five days. Degrees are awarded entirely on the marking of these, no assessments being invited from tutors. For me one nagging thought was the number of times I had been told that increasing age makes it difficult to retain information. I recalled that when pilots had studied for the Air Registration Board examinations on the various aircraft to which we were assigned the youngest had assimilated the details more easily than their elders. They would leave the examination room an hour earlier than Captains struggling to complete the answers in the allotted time.

To my great relief however I found that I could recall a sufficient knowledge of history far more successfully than I had been able to master the electrical system of a Britannia or the hydraulics of a VC-10. All went well and having obtained second class honours I dressed up once more to receive the degree from the Vice-Chancellor amid the solemn ceremony of the Sheldonian Theatre.

During those three most enjoyable years at the University I had given very little thought to any future plans beyond my final examinations. Unlike my young contemporaries I did not have to contemplate immediate employment and was content to pursue one goal at a time. It was suggested that I should offer my services as a history tutor to one or more of the tutorial colleges which proliferate in university towns including Oxford. These institutions offer individual tuition by graduates in their chosen subject to pupils aiming at university entrance in addition to 'A' levels. I was able to undertake this type of work through the kindness of my college authorities who made a room available to me to receive pupils. In addition I was asked to act as tutor to some of the college's history candidates.

I also found the time to pursue some further studies myself. I enrolled in an evening class in 'A' level Spanish. My fellow pupils were all adults and included at least one doctor who sometimes had to leave us in a hurry when her 'bleeper' sounded. Such was the enthusiasm of our teacher as well as our class that the janitor would rattle his keys at us at 9.30 pm and enquire whether we did not all have homes to return to.

Ten pleasurable years after leaving British Airways I retired from tutoring also. I continue to enjoy writing and hope that I can entertain more readers as fascinated by aviation as I am myself.